Hydraulic Machines: Fundamentals

M. A. Murtaza

Dedicated to

My Family and Friends.

PREFACE

The book, "Hydraulic Machines: Fundamentals", delves into Fundamentals, Turbines, Centrifugal Pumps, their performance evaluation and applications. It caters to the syllabus of UG (BE/BS) students in Mechanical/ Electrical engineering.

The examples and descriptions worked, enhancing the overall learning experience. The simplified language and clear explanations make this book an invariable resource for understanding Hydropower principles, applications, research and future   trends.

ACKNOWLEDGMENTS

Writing a book is a unique experience. My background includes diverse educational institutions and professional environments. My BS in Mechanical Engineering is from Harcourt Butler Technological Institute Kanpur and MS & PhD is from Motilal Nehru National Institute of Technology Allahabad.

I had academic environment of National Academy of Railways Vadodara, Manchester Business School UK, RajaRajeswari College of Engineering Bangalore, Oxford College of Engineering Bangalore, Amity School of Engineering & Technology Lucknow.

Furthermore, my international experience includes stints at Westinghouse Brakes & Signal Company Chippenham UK, and Transport Test Centre Pueblo USA.

Memberships of Institute of Science & Technology, American Society of Mechanical Engineers, Institute of Engineering & Technology, Institution of Mechanical Engineers UK, and Indian Railway service of Mechanical Engineers were part of my professional journey.

I acknowledge the insights, feedback, and encouragement which I received from my colleagues and friends in writing this book.

I also acknowledge the assistance which I received from KDP and their resources.

This book is the result of collective efforts, support, and resources of all those mentioned above.

M.A. Murtaza

(mamurtaza@hotmail.com)

Contents

Chapter 1: Basics of Hydrodynamic Machines

Prerequisites

1. Fluid is a substance which is capable of flowing, has no definite shape of its own, can conform to the shape of the containing vessel.
2. Ideal fluid is incompressible, has no viscosity and shear force.

3. Real fluid is compressible, has viscosity and shear force.

4. Viscosity is the property of a liquid due to which it resists the movement of one layer of liquid over another adjacent layer.
5. Compressibility is the property by virtue of which fluids undergo a change in volume under the action of external pressure.
6. Newtonian law of Viscosity: The shear force acting between the layers of fluid is proportional to the

difference in their velocities and the area of the plate and inversely proportional to the distance between them.

7. Momentum equation is based on the law of conservation of momentum, accordingly, the net force acting on a fluid mass is equal to the change in momentum of flow per unit time in that direction.

8. Laminar flow: There is a shear stress between fluid layers with 'No slip' at the boundary, The flow is irrotational with continuous dissipation of energy due to viscous shear. Head loss due to friction-Darcy formula is $h_f = \frac{4flV^2}{2gD}$

1.1

[Where, f = Coefficient of friction in pipe; L = Length of the pipe; D = Diameter of pipe and V = velocity of the fluid].

9. Forces present in fluid flow are Inertia force; Viscous force; Surface tension force & Gravity force.

10. Steady flow: various characteristics of following fluids such as velocity, pressure, density, temperature etc. at a point that do not change with time.

11. Assumptions made in deriving Bernoulli's equation are the fluid is ideal, The flow is steady, the flow is incompressible & the flow is irrotational.

12. Bernouillie's equation for real fluid:

$$\frac{p_1}{\rho g} + \frac{V_1^2}{2g} + Z_1 = \frac{p_2}{\rho g} + \frac{V_2^2}{2g} + Z_2$$

1.2

Where, $\frac{p}{\rho g}$ -Pressure energy. $\frac{v^2}{2g}$
=Kinetic energy. z-Datum
energy.

Impulse and Reaction Machines

In an impulse machine, there is no change in static pressure in the rotor, so that the reaction R is zero. Figure 1 shows a paddle wheel, which works on this principle.

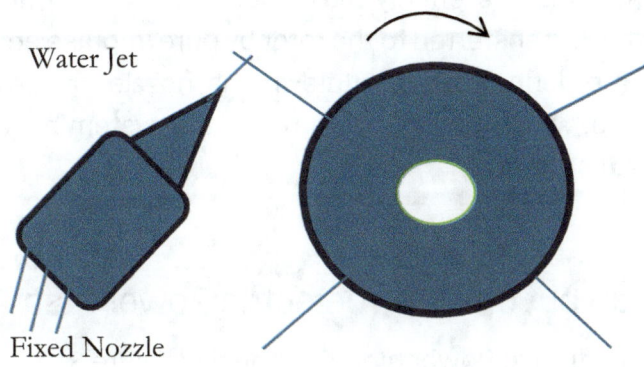

Figure 1: Paddle Wheel

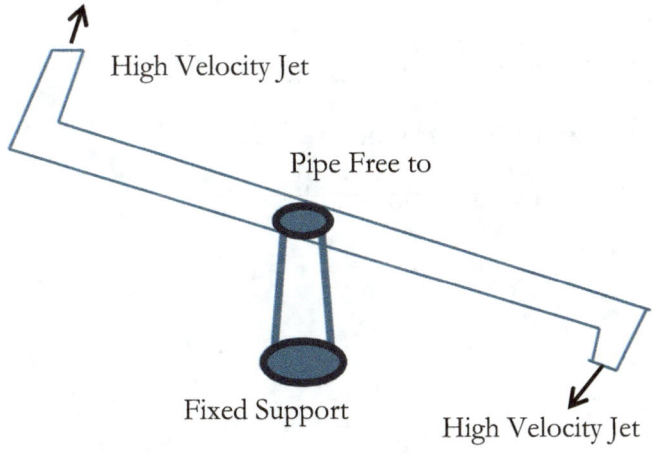

Figure 2: Lawn Sprinkler

A lawn sprinkler is an example of a reaction machine, in which water comes out at higher velocity from the rotor in the tangential direction. In lawn sprinkler rotor, the water enters at high pressure and is transformed into kinetic energy by a nozzle, which is part of the rotor itself.

In the paddle wheel, the nozzle is stationary, and its function is only to transform pressure energy into kinetic energy and finally this kinetic energy is transferred to the rotor by pure impulse action. The change in momentum of the fluid in the nozzle gives rise to a reaction force, which is not transferred to the system because the nozzle is held stationary.

General Layout of a Hydroelectric Power Plant

Layout of a hydraulic Power plant is shown in Figure 3.

H_g = Gross head

$$h_f = \frac{4fLV^2}{2g.D}$$

Where V = Velocity of flow in penstock

L = Length of penstock

D = Diameter of penstock

Net head $H = H_g - h_f$

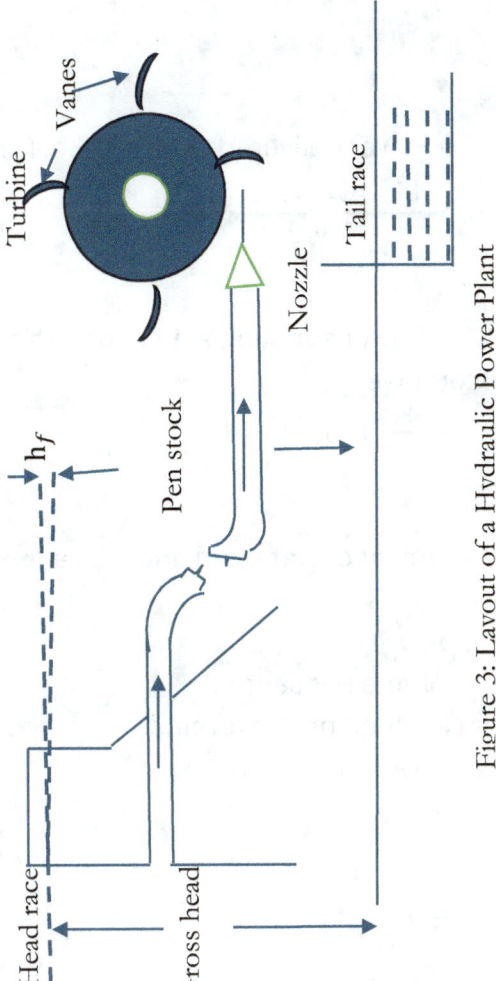

Figure 3: Layout of a Hydraulic Power Plant

Efficiencies of a Turbine

(a) Hydraulic efficiency (η_h) is defined as the ratio of power given by water to the runner of a turbine (runner is a rotating part of the turbine and on the runner, vanes are fixed) to the power supplied by the water at the inlet of turbine.

$$\eta_h = \frac{Power\ delivered\ to\ runner}{Power\ supplied\ at\ inlet} = \frac{RP}{WP}$$

9

Where, RP = Power delivered to runner i.e. runner power

$$= \frac{\frac{W}{g}[V_{\omega 1}.u_1 \pm V_{\omega 2}.u_2]}{1000} \text{ Kw}$$

WP = Power supplied at inlet of turbine and also called waterpower

$$= \frac{WH}{1000} \text{ KW}$$

Where W = Weight of water striking the vanes of the turbine per second.

$$= \rho g Q$$

Q = Volume per second.

H = Net head on the turbine.

$$WP = \frac{\rho g Q H}{1000} \text{ KW}$$

For water $\rho = 1000 \frac{kg}{m^3}$

(b) Mechanical efficiency (η_m)

The ratio of the power available at the shaft of the turbine (Known as shaft power: S.P. or brake power: B.P.) to the power delivered to the runner is defined as mechanical efficiency.

$$\eta_m = \frac{Power\ at\ the\ shaft\ of\ the\ turbine}{Power\ delivered\ by\ water\ to\ the\ runner} = \frac{S.P.}{R.P.}$$

(c) Volumetric efficiency (η_v)

The ratio of the volume of the water actually striking the runner to the volume of water supplied to the turbine is defined as volumetric efficiency.

$$\eta_v = \frac{Volume\ of\ water\ actually\ striking\ the\ runner}{Volume\ of\ water\ supplied\ to\ the\ turbine}$$

1.8

(d) Overall efficiency (η_o)

It is defined as the ratio of power available at the shaft of the turbine to the power supplied by the water at the inlet of the turbine.

$$(\eta_o) = \frac{Power\ available\ at\ the\ shaft\ of\ the\ turbine}{Power\ supplied\ at\ the\ inlet\ of\ the\ turbine} = \frac{SP}{WP}$$
$$= \frac{SP}{RP} = \frac{SP}{RP} \cdot \frac{RP}{WP}$$

1.9

Therefore,

$$\eta_o = \eta_m \times \eta_h$$

1.10

$$\eta_o = \frac{Shaft\ power\ in\ KW}{Water\ power\ in\ KW} = \frac{P}{\dfrac{\rho\,g\,Q\,H}{1000}}$$

1.11

Hydraulic Turbines

A hydraulic turbine is a prime mover that uses raw energy of flowing water and converts it into mechanical energy, in the form of rotation of its runner. This runner is directly coupled to the electric generator. Thus, electric power is generated and transmitted over long distances with the help of transmission towers.

The first hydroelectric station was started in the USA in 1882. In India it was started in 1902 in Mysore.

Hydropower is a conventional renewable source of energy which is clean, free from pollution and generally has good environmental effects. However, it has the following constraints.

i)   Large investments

ii)   The long gestation period that is the concept and development of plan & execution takes long period of time.

iii)   Increased cost of power transmission.

Hydro Power

The term hydro power means generation of mechanical power from falling water. The power then can be used for direct mechanical applications or for generating electricity. Hydropower presently accounts for 20% of the world's total electric power generation [1].

Other resources such as solar, wind, geothermal and tidal energy were also explored. These are green and renewable and have no adverse effect on the environment. These are produced continuously and therefore inexhaustible. In fact, hydro power is also an outcome of abundant solar power and water.

Possibility of using hybrids of hydro, solar and wind power are also being attempted [2, 3, 4, 6]. Among these, hydropower is the oldest and was in use during agriculture-based society [5].

The development of hydroelectricity in the 20th century required the building of large dams. These have also resulted in environmental problems due to the interference with river flow.

The current focus is on small or micro hydro power generation as they do not require massive construction. Small-scale hydro is generally a "run-of-river", with no dam or water storage. It is cost-effective and environmentally friendly energy technology [7]. Hence several "run-of-river" schemes are found on the downstream end of the large storage reservoir schemes. The large reservoir project can regulate the output of these small "run-of-river" plants. These schemes are considerably cheaper as little construction is required. [6, 8].

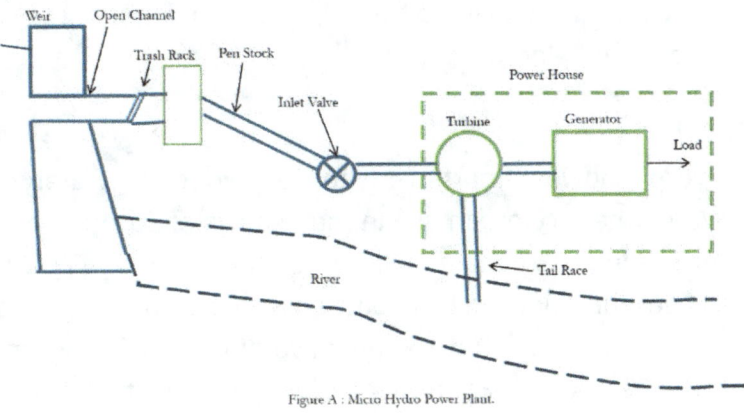

Figure A : Micro Hydro Power Plant.

Classification of Hydro Power Plants

The following table gives the classification of hydro power plants. It is based on capacity and worldwide practice [7, 10].

Table 4.1: Classification Based on Capacity

Type	Capacity
Large Hydropower Plants	> 25 MW
Small Hydropower plants (SHPPs)	2.5 MW – 25 MW
Micro Hydropower plants (MHPPs)	< 500 KW
Pico Hydropower plants	Below 10 KW

The energy conversion is through the prime mover known as hydro turbine. Hydro turbines convert water energy into mechanical energy.

History of Hydraulic Turbines

Hydro power started with a wooden waterwheel. Waterwheels were in use in many parts of Europe and Asia for some 2,000 years, mostly for milling grain [7]. During 17th century industrial revolution started and hydro power got a boost and led to the development of modern-day turbines. Benoît Fourneyron designed the first hydraulic turbine in France in 1820s and called it a hydraulic motor. Waterwheels were replaced by turbines and then the focus was to exploit hydro power for large-scale supply of electricity.

Pelton wheel (or turbine) is a high head tangential flow impulse turbine developed in 1870s by an American engineer, L.A. Pelton. In India these turbines are installed in hydro plants at Koyna, Shravathi, Joginder Nagar, Pallivasal and Pykra [8,9].

Francis turbine is a mixed flow reaction turbine developed by J. B. Francis, an American engineer, and his team at Lowell, Massachusettsin in 1848. In the following years, Francis turbine was the most widely used turbine. These turbines are installed in India in hydro plants located at Bhakra, Siva Samudaram, GanghiSagar, Hirakud and Rihand.

The growing demands of higher capacity, velocity and efficiency led to the development of the Kaplan turbine – an axial flow reaction turbine patented by an Austrian engineer, V. Kaplan in 1913. Hydro plants, in India, located at Hirakud, Tungabhadra, Kohlapur and NizamSagar use these turbines.

In impulse turbines the water energy is entirely converted into kinetic energy. The high-speed water jet enters the turbine at atmospheric pressure. Therefore, the turbine is not pressure tight. In a reaction turbine, on the other hand, the entering water jet has kinetic energy as well as pressure energy as it enters the turbine impeller. Therefore, the turbine is pressure tight. Consequently, the manufacturing cost of reaction turbines is higher. In addition to the above turbines, Turgo turbines are impulse type, and the propeller turbine is a reaction type. Pelton, Francis and propeller (a variant of Kaplan) find applications in hydropower plants. The criteria for the choice of the most suitable and efficient turbine for an application are the head and the flow rate [1, 6, 12].

Teaching Concepts of Hydraulic Machines

Hydropower is taught in the 'Advanced Fluid Mechanics' and 'Thermal & Fluid Power' courses in UG(Mech.) & UG(Elect.) engineering programs. The prerequisites of these courses are 'Basic Fluid Mechanics' and a course in 'Engineering Mathematics'. The theoretical input of the

15

course includes impact of jets, Euler's equation of hydrodynamic machines, hydraulic turbines (Pelton, Francis and Kaplan) and their performance parameters, specific speed and unit quantities, selection and governing of turbines, pumps and performance characteristics, cavitation, hydraulic transmission and hydraulics. The course is taught over 70 hours of lectures. The additional weekly laboratory sessions include performance analysis of Pelton, Francis & Kaplan turbines, centrifugal pumps, and a set up for study of cavitation, pipe friction and hydraulic control.

The taught and practical sessions are run such that theory is taught before experiments are performed in the laboratory. The normal sequence of activities includes theory classes followed by a laboratory experiment and its analysis and numerical exercises covering the taught theoretical concepts.

Performance Evaluation [Modeling, Simulation & Practical]

Modeling, simulation and practical evaluation are the three approaches for performance assessment. Modeling of hydraulic plants has been done [13] for linear and non-linear systems with elastic and non-elastic water column by using transfer function approach. With numerical techniques and increased processing speed of computers, simulation has been used for performance predictions as well as for research and development by using CFD in packages such as ANSYS/CFX [14, 15]. These approaches require validation using practical or experimental data. Mathematical modeling of the type such as hydraulic transmission and actuation system [16, 17] enables us to simulate performance by varying each hydraulic length, diameter and geometry, which is difficult in the context of a prime mover.

Tyagi [18] has verified experimentally and theoretically variations of flow rate, water level, power and angle of incidence including different liquids in Pelton turbine. Agar and Rasi [17] analyzed the data based on student experiments on Pelton turbine in laboratory. Analysis included variation of tangential force, speed, and mechanical efficiency. A variation of tangential force with speed of rotation was represented by fitted linear equation. Gudukeya et al [20] studied the effect of surface roughness, hardness of material on turbine efficiency. Efficiency also depends on the mounting of the water jet in a Pelton turbine.

Classification of Hydraulic Turbines

1. According to type of energy conversion.
   i) Impulse turbine: Energy available at inlet is kinetic energy (only). Pelton turbine is an impulse turbine.
   ii) Reaction turbine: Energy available at inlet is kinetic energy as well as pressure energy. Francis, Kaplan, Propeller turbines are examples of reaction turbines.

2. According to the direction of flow.
   i) Tangential flow turbines. In these turbines, water strikes the runner tangentially to the path of rotation. Pelton turbine is an example.
   ii) Radial flow turbine
   iii) Axial flow turbine. In these turbines the water flows parallel to the axis of the turbine shaft. Kaplan turbine
   iv) Mixed (radial & axial). In this water enters the blades radially and comes out axially, parallel to the turbine shaft. Modern Francis turbine.

3. According to the head at the inlet of the turbine.
   i)   High head turbine: 500 to 2000 meters: Pelton turbine
   ii) Medium head turbine: 60 to 300 meters: Francis turbine.
   iii) Low head turbine: 4 to 60 meters: Propeller & Kaplan turbines.
4. According to the specific speed of turbines

Specific speed: The specific speed of a turbine is defined as the speed of a geometrically similar turbine that would develop 1 KW under 1 meter head. All geometrically similar turbines (irrespective of the sizes) will have the same specific speeds when operating under same head.

Specific speed, $N_s = \dfrac{NP^{\frac{1}{2}}}{H^{\frac{5}{4}}}$

1.12

Where N is normal working speed.
      P is the power output.
      H is the net or effective head in meters.
Turbines with low specific speeds work under high head and low discharge conditions while high specific speed turbines work under low head and high discharge conditions.

i)   Low specific speed 12 to 70   Pelton turbine
ii) Medium specific speed 80 to 400 Francis turbine.
iii) High specific speed 300 to 1000   Propeller & Kaplan turbines.

Impact of Jets

The liquid comes out in the form of a jet from the outlet of a nozzle. That nozzle is fitted to a pipe, through which the liquid is flowing under pressure.

A fluid jet is a stream of fluid issuing from a nozzle with a high velocity and hence a high kinetic energy. When jet impinges on a plate or vane, it exerts a force on it (due to change in momentum). This force (hydrodynamic) can be evaluated by using Newton's second law of motion or 'impulse-momentum principle or equation'.

Force exerted by Jet on a Vertical Plate

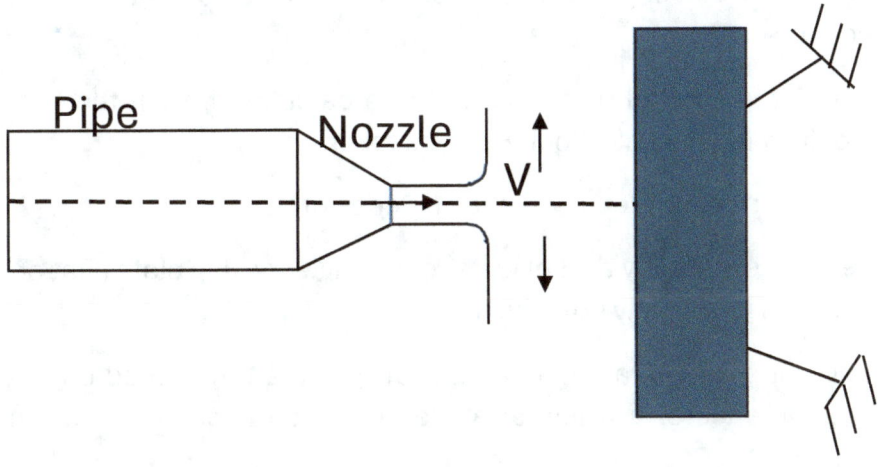

Figure 4: Force Exerted by a Jet on a Vertical Plate.

As shown in Figure 4, the jet after striking the plate will move along the plate, (which is at right angle to the plate).

The component of velocity in the direction of jet, after striking will be zero.

19

The force exerted by the jet on the plate in the direction of jet,

F_x = Rate of change of momentum in the direction of force

$$= \frac{\text{Initial momentum} - \text{Final momemtum}}{\text{Time}}$$

$$= \frac{\text{Mass x Initial Velocity} - \text{Mass x Final Velocity}}{\text{Time}}$$

$$= \frac{\text{Mass}}{\text{Time}}(\text{Initial Velocity} - \text{Final Velocity})$$

$$= \rho a V(V - 0)$$

$$= \rho a V^2$$

1.13

Where a= area of cross-section of jet, ρ= density of the fluid of jet and V is the velocity of the jet. Therefore, mass of fluid striking per second = ρaV.

If the force exerted on the jet is to be calculated, then the final velocity minus initial velocity is to be taken.

Force exerted by a Jet of water on a Series of Vanes

The force exerted by a jet of water on a single moving plate (Flat or curved) is practically not feasible.

In actual practice, a large number of plates are mounted on the circumference of a wheel at a fixed distance apart as shown in figure 5.

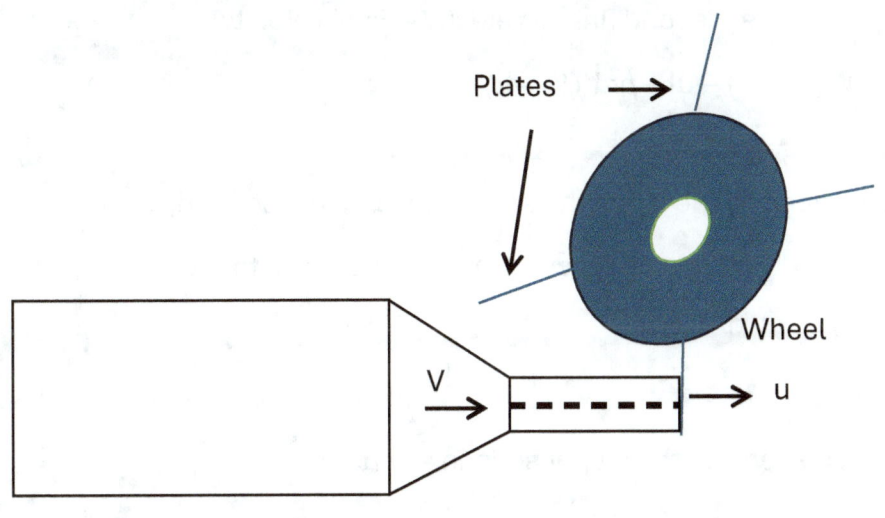

Figure 5: Jet of Water Striking a Series of Vanes.

The jet strikes a plate and due to the force exerted by the jet on the plate, the wheel starts moving and the second plate mounted on the wheel appears before the jet, which again exerts force on the second plate. Thus, each plate appears successively before the jet and jet exerts force on the plate. The wheel starts moving at a constant speed.

Mass of water striking per second, the series of plates = ρaV

The jet strikes the plate with a velocity = (V-u),

Where u = velocity of vane

After striking, the jet moves tangentially to the plate and hence the velocity component in the direction of motion of plate is equal to zero.

The force exerted by the jet in the direction of plate is equal to zero.

The force exerted by the jet in the direction of motion of plates,

F_x= Mass per second [Initial velocity – Final velocity]

$=\rho aV[(V-u)-0] = \rho aV(V-u)$

1. 14

Work done by the jet on the series of plates per second,

= Force x Distance per second in the direction of force

$=F_x.u = \rho aV(V-u).u$

1. 15

Kinetic energy of the jet per second $= \frac{1}{2}mv^2$

$= \frac{1}{2}\rho aV.V^2 = \frac{1}{2}\rho aV^3$

1. 16

Efficiency $\eta = \dfrac{Work\ done\ per\ second}{(Kinetic\ energy\ per\ second)}$

$= \dfrac{[\rho aV(V-u).u]}{\left[\frac{1}{2}\rho aV^3\right]}$

$=[2u(V-u)]/[V^2]$

1.17

Condition for maximum efficiency for a given jet velocity.

$$\frac{d}{du}\eta = 0 = \frac{d}{du}\left[\frac{[2u(V-u)]}{V^2}\right] = 0$$

Or V= 2u or u $=\dfrac{V}{2}$

Therefore,

$\eta_{max} = 2u[2u-u]/(2u)^2 = \frac{1}{2} = 0.5 = 50\ \%$

1.18

Force exerted on a Series of Radial Curved Vanes

The radius of the vane at inlet and outlet is different for a radial curved vane. Therefore, the tangential velocities of the radial vane at inlet and outlet will not be equal.

The wheel, having a series of radial curved vanes, starts rotating at a constant angular speed due to the impact of jet.

Let R_1 = Radius of wheel at inlet of the vane

R_2 = Radius of wheel at outlet of vane

ω = Angular speed of the wheel

The velocity triangle at the inlet and the outlet are drawn as shown in the figure 6.

The mass of water striking per second for series of vanes,

= $\rho a V_1$, Where a = Area of jet & V_1 = Velocity of jet.

Momentum of water striking the vanes in the tangential direction per second at inlet,

= Mass of water striking per second x Component of V_1 in the tangential direction.

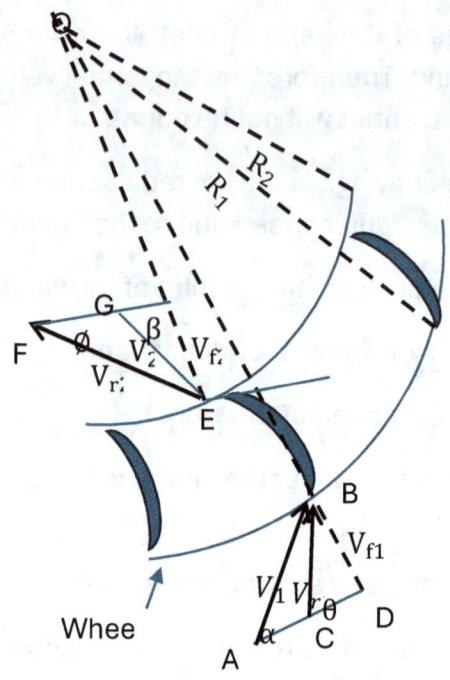

Figure 6: Series of Radial Curved Vanes mounted on a

$= \rho a V_1 \times V_{\omega 1}$ [Because component of V_1 in tangential direction = $V_1. \cos \alpha = V_{\omega 1}$

1.19

Similarly, momentum of water at the outlet per second,

$= \rho a V_1 \times$ Component of V_2 in tangential direction,

$= \rho a V_1 \times (- V_2 \cos \beta) = - \rho a V_1. V_{\omega 2}$; Because $V_2 \cos \beta = V_{\omega 2}$

1.20

-ve sign is taken as the velocity V_2 at outlet is in the opposite direction.

Now angular momentum per second at inlet = Momentum at inlet x Radius at inlet

$= \rho a V_1 \times V_{\omega 1} \times R_1$

1.21

Similarly angular momentum per second at outlet = $-\rho a V_1 . V_{\omega 2} \times R_2$

1.22

Therefore, the Torque exerted by the water on the wheel,

T = Rate of change of angular momentum,

$= \rho a V_1 \times V_{\omega 1} \times R_1 - (-\rho a V_1 . V_{\omega 2} \times R_2)$

$= \rho a V_1 [V_{\omega 1} \times R_1 + V_{\omega 2} \times R_2]$

1.23

Work done per second on the wheel,

$=$ Torque x Angular velocity $= T \times \omega$

$= \rho a V_1 [V_{\omega 1} \times R_1 + V_{\omega 2} \times R_2] \times \omega,$

$= \rho a V_1 [V_{\omega 1} \times u_1 + V_{\omega 2} \times u_2]$

1.24

Where $u_1 = \omega R_1$ and $u_2 = \omega R_2$

If the angle β in the figure is an obtuse angle, then work done per second will be given as

$= \rho a V_1 [V_{\omega 1} \times u_1 - V_{\omega 2} \times u_2]$

1.25

Therefore, the general expression for work done per second on the wheel,

$$= \rho a V_1 \left[V_{\omega 1} \times u_1 +/- V_{\omega 2} \times u_2 \right]$$

$$1.26$$

If the discharge is radial at the outlet $\beta = 90°$, then work done becomes,

$$= \rho a V_1 \left[V_{\omega 1} \times u_1 \right]$$

$$1.27$$

Efficiency of the Radial Curved Vane

The work done per second on the wheel is the output of the system, whereas the kinetic energy per second of the jet is the input of the system. Hence the efficiency of the

$$\eta = \frac{\text{Work done per second}}{\text{Kinetic energy per second}}$$

$$= \rho a V_1 \left[V_{\omega 1} \times u_1 +/- V_{\omega 2} \times u_2 \right] / \left[\left(\frac{1}{2} \right) (\rho a V_1) V_1^2 \right]$$

$$= \frac{2}{V_1^2} \left[V_{\omega 1} \cdot u_1 +/- V_{\omega 2} \cdot u_2 \right]$$

$$1.28$$

If there is no loss of energy when water is flowing over the vanes, the work done on the wheel per second is also equal to the change in kinetic energy of jet per second. Therefore,

Work done per second on wheel = change in kinetic energy per second of the jet

$$= \frac{1}{2} m (V_1^2 - V_2^2) = \frac{1}{2} (\rho a V_1) (V_1^2 - V_2^2)$$

$$1.29$$

Hence efficiency $= \left[\frac{1}{2}(\rho a V_1)(V_1^2 - V_2^2)\right]/\left[\frac{1}{2}(\rho a V_1)(V_1^2)\right] = \left[1 - \left(\frac{V_2^2}{V_1^2}\right)\right]$

1.30

From the above equation, it is clear that for a given initial velocity of the jet (i.e. V_1), the efficiency will be maximum, when V_2 is minimum. But V_2 cannot be zero as in that case incoming jet will not come out of the vane.

Equation also gives the efficiency of the system. From this it is clear that the efficiency will be maximum when $V_{\omega 2}$ is added to $V_{\omega 1}$. This is only possible if β is an acute angle (less than 90°). Also, for maximum efficiency $V_{\omega 2}$ should also be maximum. This is only possible if $\beta = 0$. In that case $V_{\omega 2} = V_2$, angle ϕ will be zero. But in actual practice ϕ cannot be zero. Hence for the maximum efficiency, ϕ must be minimum.

Jet Propulsion

Jet Propulsion means the propulsion or movement of bodies such as ships, aircraft, and rocket with the help of jet.

According to Newton's third law of motion, to every action there is an equal and opposite reaction. The orifices are on the bodies, the reaction of the jet coming out from their orifice is used to move them.

The magnitude of force exerted is equal to the 'action of the jet'. This force which is acting on the orifice or nozzle in the opposite direction is called the reaction of the jet'. If the body in which the orifice or nozzle is fitted is free to move, the body will start moving in the direction opposite to the jet.

Jet Propulsion of a Tank with an Orifice

Consider a large tank fitted with an orifice on one of its sides as shown in Figure 7.

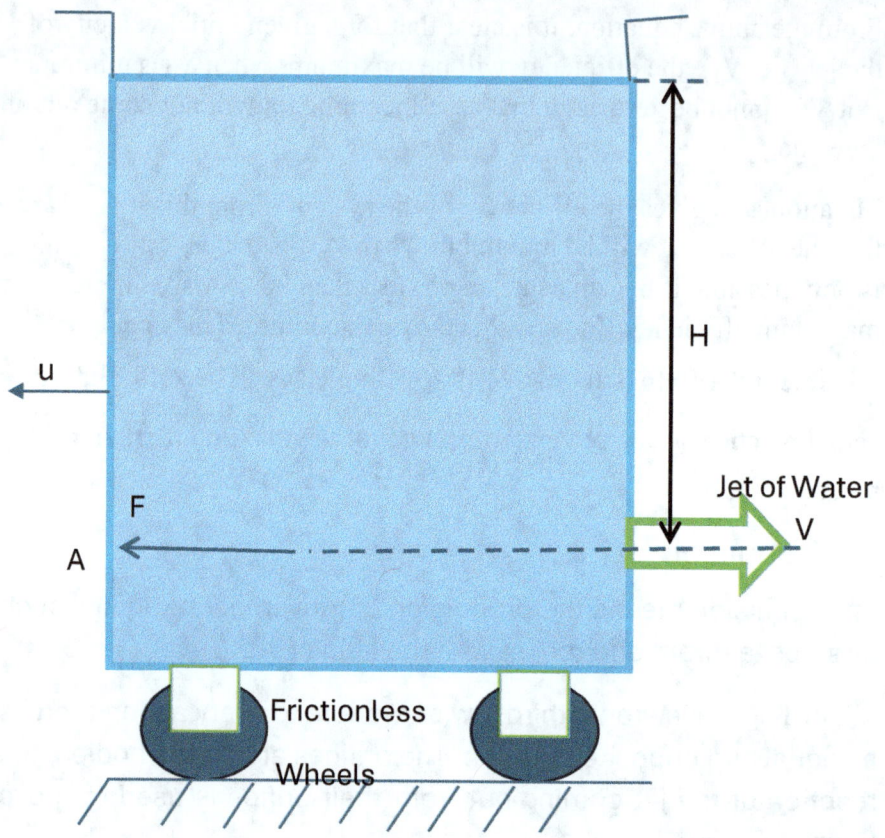

Figure 7: Jet Propulsion of a Tank with an Orifice.

H = Constant head of water in tank from the centre of orifice.

a = area of orifice.

28

V = velocity of jet of water.

$$V = C_v(2gH)^{\frac{1}{2}}$$

1.31

C_v = Coefficient of velocity of orifice.

Mass of water coming out from the orifice per second = ρ x Volume per second. = ρaV.

Force acting on water is equal to the rate of change of momentum.

Initial velocity of water in the tank is zero and final velocity of water when it comes out in the form of jet is equal to V.

Therefore F = ρaV. [V-0] = $\rho a V^2$

1.32

This jet of water will exert a force on the tank which is equal to F but in opposite in direction as shown in Figure. The force will be acting at A, the point on the tank in the horizontal line of the centre of the orifice. If the tank is free to move or tank is fitted with frictionless wheel, it will start moving with some velocity say u, in the direction opposite to the direction of the jet.

When the tank starts moving, the velocity of the jet with which it comes out of the orifice will not be equal to V but will be equal to the relative velocity of the jet with respect to tank.

$V_r = V - (-u) = V + u$

1.33

Hence when the tank is moving, the velocity with which jet comes out from the orifice is (V + u).

Mass of water coming out from the orifice per second = ρ x a x Velocity with which water comes out.

= ρ a V_r = ρ a V_r = ρ a (V + u)

29

Force exerted on tank is given by

$$F_x = \rho a (V + u) [(V + u) - u] = \rho a (V + u) [(V + u) - u] = \rho a (V + u) V$$

Work done per second $= F_x . u = \rho a (V + u) V (V.u)$

Efficiency of propulsion is given by

$$\eta = \frac{Work\ done\ per\ second}{Kinetic\ energy\ of\ the\ jet\ per\ second}$$

$$= \frac{[\rho a (V + u) V (V.u)]}{\left(\frac{1}{2}\right)\rho a (V + u) V ((V+u)^2)}$$

$$= \frac{[2Vu]}{(V+u)^2}$$

Condition for maximum efficiency, $\frac{d\eta}{du} = 0$

It gives u = V

Therefore $\eta_{max} = \frac{1}{2} = 0.5$ or 50%

Jet Propulsion of Ships

The jet propulsion principle is used to drive a ship through water. A ship carries centrifugal pumps and nozzles. Water is drawn by the pump from the surroundings and is discharged through nozzle at the back (also called stern) of the ship. It exerts a propulsive force in the opposite direction on the ship.

The water from the surrounding sea by the centrifugal pump is taken in the following two ways.

Inlet Orifices are at Right Angles (Amid Ship Figure 8)

Let V_r = Relative velocity of jet with respect to ship = V + u

Propulsion force exerted F = Mass of water issuing per second x Change of velocity

$$= \rho a (V + u)[V_r - u] = \rho a (V + u). V$$

1.38

Work done per second = F x u

$$= \rho a (V + u). V.u$$

1.39

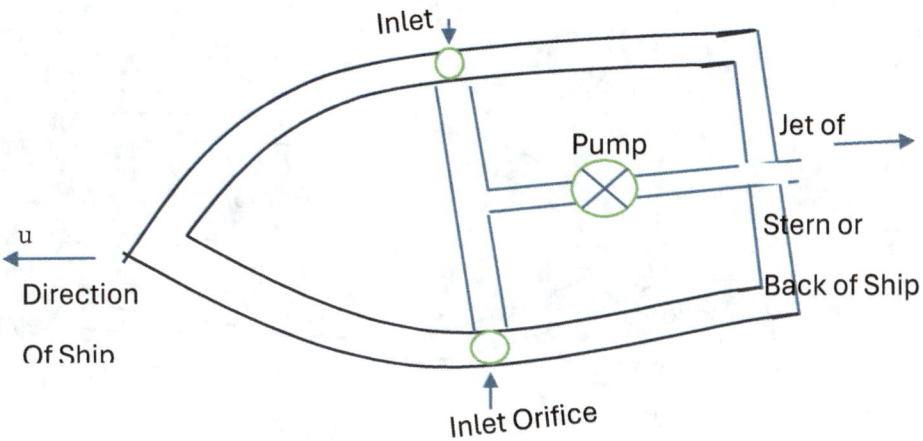

Figure 8: Inlet Orifices are at right angles (AMID SHIP)

Energy supplied by the jet (Figure 8) $= \frac{1}{2}$
(*Mass of water supplied per second*). (V_r^2)

$$= \frac{1}{2}(\rho a V_r)(V_r^2)$$

1.40

Therefore, Kinetic energy supplied by jet $= \frac{1}{2}\rho a (V + u)[(V + u]^2]$

1.41

Therefore, efficiency of propulsion $\eta = \dfrac{Work\ done\ per\ second\ by\ jet}{Energy\ supplied\ by\ jet}$

31

$$= (\rho\, a\, (V + u).\, V.u)/[\tfrac{1}{2}\rho\, a\, (V + u)[(V + u]^2]$$

$$= \frac{2Vu}{(V+2u)^2}$$

1.42

Inlet Orifices Facing the Direction of Ship

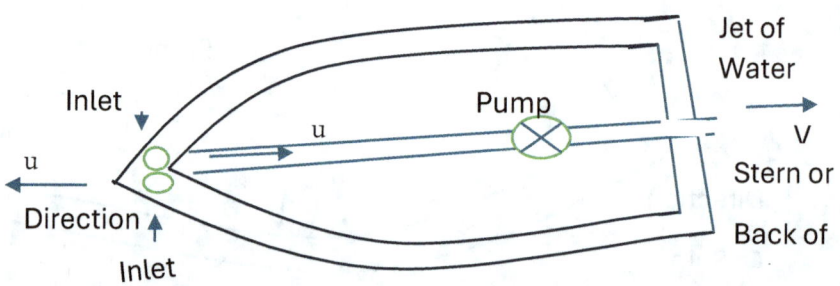

Figure 9: Inlet Orifices facing the direction of ship.

There is no change in the expression of work done. However energy supplied by the jet (Figure 9), water enters ship with a velocity equal to that of ship $u = \frac{1}{2}$

(Mass of water supplied per second). $(V_r^2 - u^2)$

$$= \tfrac{1}{2}(\rho\, a\, V_r)(V_r^2 - u^2)$$

1.43

Therefore, Kinetic energy supplied by jet $= \frac{1}{2} \rho a (V + u)[(V + u]^2 - u^2]$

1.44

Therefore, efficiency of propulsion $\eta = \dfrac{Work\ done\ per\ second\ by\ jet}{Energy\ supplied\ by\ jet}$

$$= (\rho a (V + u).V.u)/[\frac{1}{2}\rho a (V + u)[(V + u]^2 - u^2]$$

$$= \frac{2u}{V + 2u}$$

1.45

Worked Examples

Problem 1:

A jet of water having a velocity of 35 m/s, impinges on a series of vanes moving with a velocity of 20 m/s. The jet makes an angle of 30° to the direction of motion of vanes when entering and leaves at an angle of 120°. Draw triangles of velocities at inlet and outlet.

Find: (a) The angle of vane tips so that water enters and leaves without shock.

     (b) the work done per unit weight of water entering the vanes and

     (c) Efficiency.

Solution:

Velocity of jet $V_1 = 35\ m/s$;

Velocity of vane $u_1 = u_2 = 20\ m/s$

Angle of jet at inlet, $\alpha = 30^o$

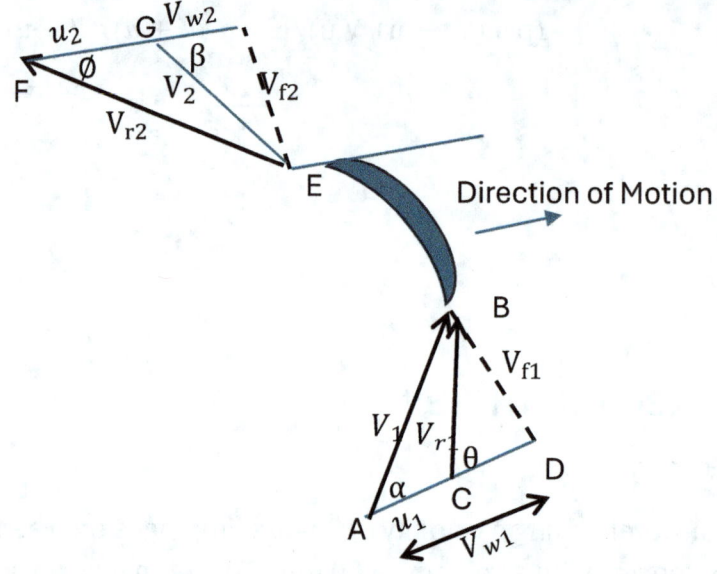

Figure: Problem 1.

Figure shows inlet and outlet velocity triangles.

$$Angle\ \beta = 180^{o} - 120^{o} = 60^{o}$$

(a) Angle of vane tips,

From inlet velocity triangle $V_{w1} = V_1 Cos\alpha = 35.Cos30^o$
$$= 30.31\ m/s$$
$$V_{f1} = V_1 Sin\alpha = 35.Sin30^o = 17.50\ m/s$$
$$tan\theta = \frac{V_{f1}}{V_{w1} - u_1} = 1.697$$

$V_{r1} = \frac{V_{f1}}{Sin\theta} = 20.25$ m/s

Taking $V_{r2} = V_{r1}$ (Condition of negligible losses).

From outlet velocity triangle.

$$V_{r2} = 20.25 = \frac{u_2 Sin120^o}{Sin(60^o - \emptyset)}$$

It will give value of $\emptyset$; $\emptyset = 1.25^o$

(b) Work done per unit weight of water entering

$$= \frac{1}{g}(V_{w1} + V_{w2}).u_1$$
$V_{w2} = V_{r2} Cos\emptyset - u_2 = 0.24$ m/s
Work done per unit weight=62.28 Nm/N

(c) Efficiency= $\frac{62.28}{\frac{v_1^2}{2g}} = 99.74\%$

Problem 2:

A jet of water of 50 mm moving with a velocity of 25 m/s impinges on a fixed curved plate tangentially at one end at an angle of 30° to the horizontal. Calculate the resultant force of the jet on the plate if the jet is deflected through an angle of 50°. Take g= $10m^2/s$.

Solution: Area of cross section of jet= $\frac{\pi}{4}(0.05)^2 m^2$; Velocity = 25 m/s

At inlet direction of jet is 30° with horizontal, therefore direction at outlet =$30^o + 50^o = 80^o$ with the horizontal.

Therefore, $F_x = \rho a V[V Cos 30 - V Cos 80] = 849.7 N$

, $F_y = \rho a V[V Sin 30 - V Sin 80] = -594.9 N$; Means its direction is downwards.

Therefore, Resultant force $F_R = \sqrt{F_x^2 + F_y^2} = 1037 N$

Direction of resultant, $tan\alpha = \dfrac{F_y}{F_x} = 0.7$

Therefore, $\alpha = 35^o$

Problem 3:

The head of water from the centre of the orifice which is fitted to one side of tank is maintained at 2 m of water. The tank is not allowed to move, and the diameter of orifice is 100 mm. Find the force exerted by the jet of water on the tank. Take $C_v = 0.97$.

Solution: H= 2 m; d= 100 mm =0.1 m,

Area of cross-section of jet $= \dfrac{\pi}{4}d^2 = 0.007854 \, m^2$

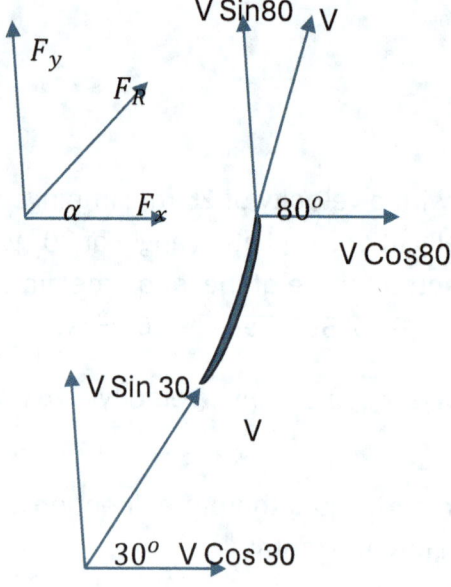

Figure: Problem 3.

36

Velocity of jet $= C_v\sqrt{2gH} = 6.07\ m/s$

Force exerted by the jet $= \rho aV(V - 0) = 1000.(0.007854).(6.07)^2$
$= 289.3\ N$

Problem 4:

If, in the above problem, tank is fitted with frictionless wheels and allowed to move, determine

(i)   Propelling force on tank,
(ii) Work done by the propelling force per second, and efficiency of propulsion.
      The tank is moving with a velocity of 2 m/s.
      Solution:
      Velocity of tank $u = 2\ m/s$
      Propelling force $F_x = \rho a(V + u)[(V + u) - u]$

$$F_x = 384.65\ N$$
Therefore, work done $= F_x.u = 384.65 \times 2 = 769.3$ Nm/s

Efficiency of propulsion $= \dfrac{\frac{Work\ done}{second}}{\frac{Kinetic\ energy}{second}} = \dfrac{2Vu}{(V+u)^2} = 0.3728 =$
37.28%

Problem 5:
Find the propelling force acting on a ship which takes water through inlet orifices which are at right angles to the direction of motion of ship, and discharges at the back through orifices having effective areas of $0.04\ m^2$. The water flows at the rate of 1000 litres/s and ship is moving with a velocity of 8 m/s.

Solution:
Velocity of jet relative to water $= \dfrac{Q}{a}$

$$V_r = V + u = \dfrac{Q}{a}; Absolute\ velocity = \dfrac{Q}{a} - u$$

37

$Propelling\ force = \rho a(V + u).V = 1000x.04x(17+8)x17$
= 16999.94 N

Problem 6:

The water jet in a jet-propelled boat is drawn through inlet openings facing the direction of motion of the ship. The boat is moving in seawater at a speed of 30 km/hour. The absolute velocity of the jet of water discharged at the back is 20 m/s and the area of jet of water is 0.03 m^2. Find the propelling force and the efficiency of propulsion.

Solution:

Speed of boat $u = 30\dfrac{km}{hour} = 8.33\ m/s$.

Absolute velocity of jet $V = 20\ m/s$

Area of jet $a = 0.03\ m^2$

(i) Propelling force $F = \rho a(V + u)V =$

$\qquad [1000x.03x(20 + 8.33)x20] = 16997.98\ N$

(ii) Efficiency of propulsion η=

$\qquad \dfrac{Work\ done\ per\ second\ by\ jet}{Energy\ supplied} =$

$\eta = [\rho a(V + u).V.u]/[\tfrac{1}{2}\rho a(V + u)((V + u)^2 - u^2)]$

$= \dfrac{2u}{V + 2u} = 0.4544 = 45.44\%$

Chapter 2. Hydraulic Turbines

1. Pelton Wheel or Turbine

The Pelton wheel or turbine is a tangential flow impulse turbine. Thus, energy available at the inlet is in the form of kinetic energy. There is no change in pressure. It remains atmospheric at the inlet as well as at outlet of the turbine. The water enters the bucket along the tangent of the runner.

This turbine is used for high heads and is named after L. A. Pelton, an American engineer.

These turbines are installed in Koyna hydroelectric Project (Maharashtra India), Mahatma Gandhi hydroelectric Project (Karnataka India), Pallivakal Kerala, Pykara Tamil Nadu India, Maneli Himanchal Pradesh India.

Figure 3 shows the layout of a hydraulic power plant in which turbine used is Pelton wheel. The water from the reservoir flows through the penstocks at the outlet of which a nozzle is fitted. The nozzle increases the kinetic energy of the water flowing through the penstock. At the outlet of the nozzle, the water comes out in the form of a jet and strikes the bucket (vanes) of the runner.

Main parts of Pelton turbine:

1. Nozzle and flow regulation arrangement.
2. Runner with buckets.
3. Casing.
4. Braking Jet.

1.1 Velocity triangles and work done for Pelton Turbine.

The jet of water from nozzle strikes the bucket at the splitter, which splits up the jet into two parts.

These parts of the jet, glides over the inner surfaces and come out at the outer edge.

Figure 10, shows the section of the bucket at z-z section. The splitter is the inlet tip and outer edge of the bucket is the outlet tip of the bucket.

The inlet velocity triangle is drawn at the outer edge of the bucket.

Let H = net head acting on the Pelton wheel = $H_g - H_f$

2.1

Where H_g is the gross head and H_f is the friction head

$$H_f = \frac{4flV^2}{D^*.2g}$$

2.2

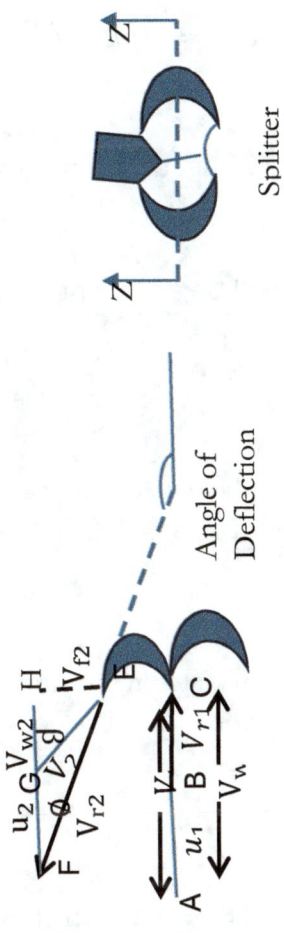

Figure 10: Velocity triangles of a Pelton Wheel.

41

Where D^* = diameter of penstock

       N = speed of wheel in rpm

       D = diameter of wheel

       d = diameter of jet

Then velocity of jet at inlet $V_1 = (2gH)^{1/2}$

$$2.3$$

Tangential velocity of the wheel u $= u_1 = u_2 = \dfrac{\pi DN}{60}$

$$2.4$$

Velocity triangle at inlet will be a straight line.

Where, $V_{r1} = V_1 - u_1 = V_1 - u$

Also, $V_{w1} = V_1, \propto = 0^o, \theta = 0^o$

$$2..5$$

From the velocity triangle at the outlet, we have

$$V_{r2} = V_{r1} \ and \ V_{w2} = V_{r2} Cos\varphi - u_2$$

$$2.6$$

The force exerted by the jet of water in the direction of motion = Mass of water striking per second [Initial velocity – final velocity in the direction of jet]

$= \rho a V_1 [V_{w1} + V_{w2}]$

$$2.7$$

Now work done by the jet on the runner per second $= F_x.u = \rho a V_1$ $[V_{w1} + V_{w2}].u \ Nm/sec$

$$2.8$$

Power given to the runner by the jet $= \dfrac{\rho a V_1 [V_{w1} + V_{w2}].u}{1000} Kw$

$$2.9$$

Work done per second per unit weight of water striking per second,

$$= \frac{\rho a V_1[V_{w1} + V_{w2}].u}{weight\ of\ water\ striking/s} = \frac{\rho a V_1[V_{w1} + V_{w2}].u}{\rho a V_1 g .u} = \left(\frac{1}{g}\right)[V_{w1} + V_{w2}]$$

2.10

The energy supplied to the jet at inlet is in the form of kinetic energy and is equal to $\left(\frac{1}{2}\right)mV_1^2$

2.11

Therefore, kinetic energy of jet per second $= \frac{1}{2}(\rho a V_1).V_1^2$

2.12

Therefore, Hydraulic efficiency $\eta_h = \dfrac{Work\ done\ per\ second}{Kinetic\ energy\ of\ jet\ per\ second}$
$$= \frac{\rho a V_1[V_{w1} + V_{w2}].u}{\frac{1}{2}(\rho a V_1).V_1^2}$$

2.13

$$= \frac{[V_{w1} + V_{w2}].u}{\frac{1}{2}.V_1^2} = \frac{[2[V_{w1} + V_{w2}].u]}{V_1^2}$$

2.14

Now at inlet: Velocities of whirl and flow are, $V_{w1}\ and\ V_{f1}$. These are components of velocity of jet V_1 in the direction of motion and perpendicular to the direction of the vane respectively.

$$V_{r1} = V_1 - u_1 = V_1 - u\ ; V_{w1} = V_1$$

2.15

At outlet, we have: $V_{r2} = k.V_{r1},$

2.16

Where k is the blade friction coefficient, slightly less than unity. Ideally when bucket surface are perfectly smooth and energy losses due to impact at splitter are neglected, k = 1,

Then $V_{r2} = V_{r1} = (V_1 - u)$

$$V_{w2} = V_{r2}Cos\varphi - u_2 = (V_1 - u)Cos\varphi - u$$

2.17

Substituting the values of V_{w1} and V_{w2} in equation above,

We get; $\eta_h = \dfrac{2[V_1 + (V_1 - u)Cos\varphi - u].u}{V_1^2}$

2.18

$$= \dfrac{[2(V_1 - u)(1 + Cos\varphi)u]}{V_1^2}$$

2.19

The efficiency will be maximum for a given value of V_1;

Therefore $\dfrac{d}{du}(\eta_h) = 0$

$$\dfrac{1 + Cos\varphi}{V_1^2}\dfrac{d}{du}(2uV_1 - 2u^2) = 0$$

2.20

As, $\dfrac{1 + Cos\varphi}{V_1^2} \neq 0$, Therefore

$2V_1 - 4u = 0$ *or* u $= \dfrac{V_1}{2}$

2.21

Equation 1.21 states that hydraulic efficiency of a Pelton wheel will be maximum, when the velocity of the wheel is half of the velocity of water jet at inlet.

Substituting the value of $u = \dfrac{V_1}{2}$ in equation1.19

Maximum; $\eta_h = \left[\dfrac{2\left(V_1 - \frac{V_1}{2}\right)(1 + Cos\varphi)V_1}{2V_1^2}\right] = (1 + Cos\varphi)$

Points to be remembered for Pelton Turbine:

(i)   The velocity of the jet at inlet is given by
$$V_1 = C_v(2gH)^{0.5}$$

Where C_v is coefficient of velocity = 0.98 or 0.99.

H = net head on turbine.

(ii)   Velocity of wheel (u) is given by

$$u = \emptyset(2gH)^{0.5}$$

Where $\emptyset$ is the speed ratio and varies from 0.43 to 0.48.

(iii)   The angle of deflection of the jet through buckets is taken 165°, if no angle of deflection is given.

(iv)   Mean diameter or pitch diameter D of the Pelton wheel is given by

$$u = \frac{\pi DN}{60} \ or \ D = \frac{60u}{\pi N}$$

(v)   Jet ratio. It is defined as the ratio of the pitch diameter (D) of Pelton wheel to the diameter of the jet (d). It is denoted by m and is given as

$$m = \frac{D}{d} = 12 \text{ for most of the cases.}$$

(vi)   Number of buckets on a runner is given by

$$z = 15 + \frac{D}{2d} = 15 + 0.5m$$

(vii)   Number of jets
It is obtained by dividing the total rate of flow through the turbine by the rate of flow of water through a single jet.

Radial Flow Reaction Turbines:

The water flows in radial direction in radial flow turbines. It is from outwards to inwards in inward radial flow turbine. It is from inward to outward in outward radial flow turbines.

The water at the inlet of the turbine posses' kinetic energy as well as pressure energy. These are reaction turbines. As the water flows through the runner, a part of pressure energy goes on changing into kinetic energy. Thus, the water through the runner is under pressure, therefore the runner is completely enclosed in an air-tight casing and the runner is always full of water.

Main Parts of a Radial Flow Reaction Turbine:
1. Casing
2. Guide Mechanism
3. Runner and
4. Draft-Tube.

Francis Turbine:

Francis Turbine was designed by J.B. Francis, an American engineer. Initially it was an inward radial flow turbine.

Modern Francis turbine is a mixed flow type, as water enters radially at the inlet and leaves at the outlet in axial direction.

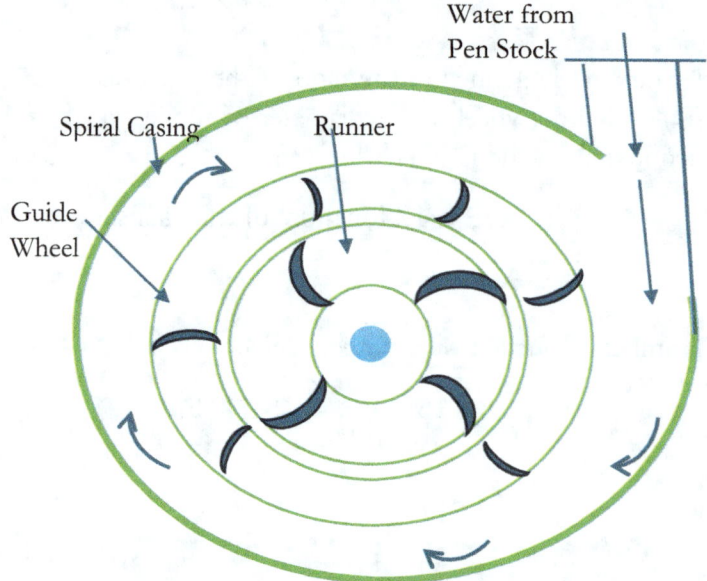

Figure 11: Main Parts of a Radial Reaction Turbine.

Velocity Triangles and Work Done on the Runner:

From the velocity triangles, the work done by the water on the runner, horsepower and efficiency of the turbine can be obtained.

The work done per second on the runner by water is given by equation (1.16) as follows

$$= \rho a V_1 [V_{w1}.u_1 \pm V_{w2}.u_2]$$

Also, $a.V_1 = Q$

The work done per second on the runner by water $= \rho Q [V_{w1}.u_1 \pm V_{w2}.u_2]$

2.28

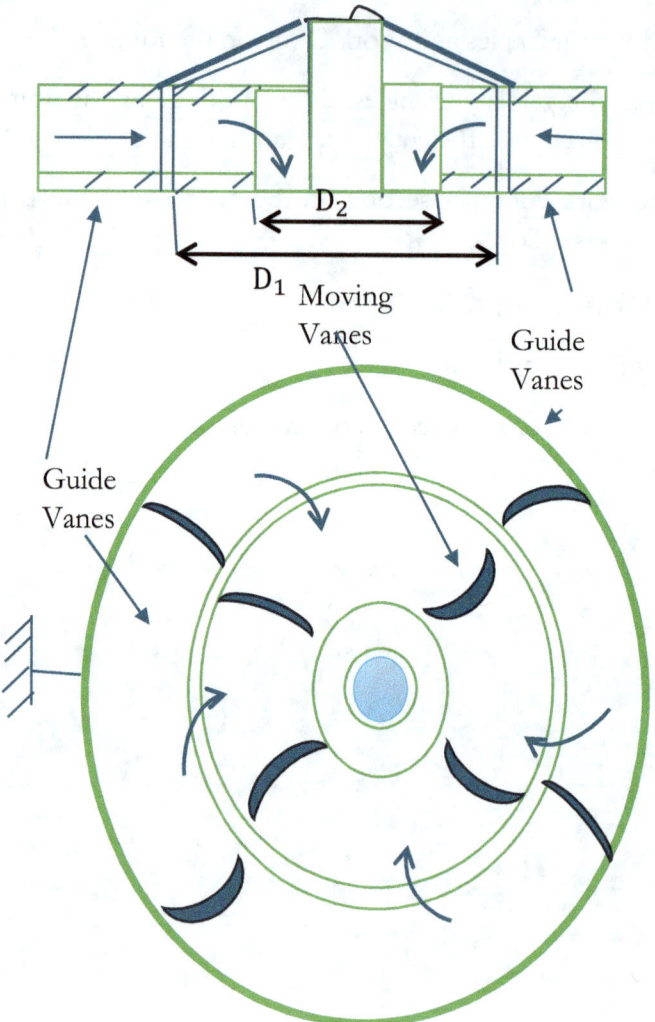

D_2

D_1 Moving
Vanes

Guide
Vanes

Guide
Vanes

Figure 12: Main Parts of an Inward Radial Flow Turbine.

The equation (2.28) also represents the energy transfer per second to the runner.

Where, V_{w1} = Velocity of whirl at inlet.

$\qquad V_{w2}$ = Velocity of whirl at outlet.

u_1 = Tangential velocity of wheel at inlet = $\frac{\pi D_1 N}{60}$, Where D_1 is outer diameter of runner.

u_2 = Tangential velocity of wheel at outlet = $\frac{\pi D_2 N}{60}$, Where D_2 is outer diameter of runner.

N is the speed the turbine in rpm.

The work done per second per unit weight of water per second,

$$= \frac{Work\ done\ per\ second}{Weight\ of\ water\ striking\ second}$$

$$= \frac{\rho Q[V_{w1}.u_1 \pm V_{w2}.u_2]}{\rho Q g} = \frac{1}{g}[V_{w1}.u_1 \pm V_{w2}.u_2]$$

$$2.29$$

The equation (2.29) represents the energy transfer per unit weight per second to the runner. This equation is known as Euler's Equation of Hydrodynamic Machines. This equation was given by Swiss scientist L. Euler.

In equation (2.29), positive sign is taken if angle β is an acute angle. If β is an obtuse angle, then negative sign is taken.

If $\beta = 90^o$, then V_{w2} = 0 and work done per second per unit weight of water striking per second becomes = $\frac{1}{g}[V_{w1}.u_1]$

$$2.30$$

Hydraulic efficiency $\eta_h = \frac{Power\ delivered\ to\ the\ runner}{Power\ supplied\ at\ inlet}$

$$2.31$$

$$= \frac{(RP)}{WP}$$

$$\eta_h = \frac{[V_{w1}.u_1]}{gH}$$

2.32

Degree of Reaction:

Degree of reaction is defined as the ratio of pressure energy change inside a runner to the total energy change inside the runner. It is represented by 'R'.

Hence mathematically, it can be written as,

$$R = \frac{Change\ of\ pressure\ energy\ inside\ the\ runner}{Change\ of\ total\ energy\ inside\ the\ runner}$$

Let H_e = Change of total energy per unit weight inside the runner. Then

$$H_e = \frac{1}{g}[V_{w1}.u_1 \pm V_{w2}.u_2]$$

2.33

Let us find the values of $V_{w1}.u_1$ and $V_{w2}.u_2$ from the velocity triangles.

$$V_{w1} = u_1 + V_{r1}Cos\theta = u_1 + (V_{r1}^2 - V_{f1}^2)^{0.5}$$

$$= u_1 + [V_{r1}^2 - (V_1^2 - V_{w1}^2)]^{0.5}$$

$$(V_{w1} - u_1) = [V_{r1}^2 - (V_1^2 - V_{w1}^2)]^{0.5}$$

Also,

$$(V_{w1} - u_1)^2 = [V_{r1}^2 - (V_1^2 - V_{w1}^2)]$$

$$V_{w1}^2 + u_1^2 + 2V_{w1}.u_1 = [V_{r1}^2 - (V_1^2 - V_{w1}^2)]$$

Therefore,

$$V_{w1}.u_1 = (\frac{1}{2})[u_1^2 - V_{r1}^2 + V_1^2]$$

2.34

Similarly,

$$V_{w2}.u_2 = (\frac{1}{2})[V_{r2}^2 - V_2^2 - u_2^2]$$

2.35

On substituting these values in equation (2.33), we get

$$H_e = \frac{V_1^2 - V_2^2}{2g} + \frac{u_1^2 - u_2^2}{2g} + \frac{V_{r2}^2 - V_{r1}^2}{2g}$$

2.36

The first term represents the change in kinetic energy of the fluid per unit weight and remaining two terms represent the change in pressure energy inside the runner.

Therefore, $R = \dfrac{\textit{Change in pressure energy inside the runner per unit weight}}{\textit{Change of total energy inside the runner per unit weight}}$

$$= \frac{[(u_1^2 - u_2^2) + (V_{r2}^2 - V_{r2}^2)]}{[(V_1^2 - V_2^2) + (u_1^2 - u_2^2) + (V_{r2}^2 - V_{r2}^2)]}$$

$$= 1 - \frac{[(V_1^2 - V_2^2)]}{[(V_1^2 - V_2^2) + (u_1^2 - u_2^2) + (V_{r2}^2 - V_{r2}^2)]}$$

$$= 1 - \frac{(V_1^2 - V_2^2)}{2gH_e}$$

2.37

(i) Values of R for Pelton Turbine:

$$u_1 = u_2 \ and \ V_{r2} = V_{r1}$$

Therefore R $= 1 - \left(\frac{(V_1^2 - V_2^2)}{(V_1^2 - V_2^2)}\right) = 1 - 1 = 0$

(ii) Values of R for an Actual Reaction Turbine:

For actual reaction turbine $\beta = 90^o$

Therefore $V_{w2} = 0$, Also $V_2 = V_{f2}$.

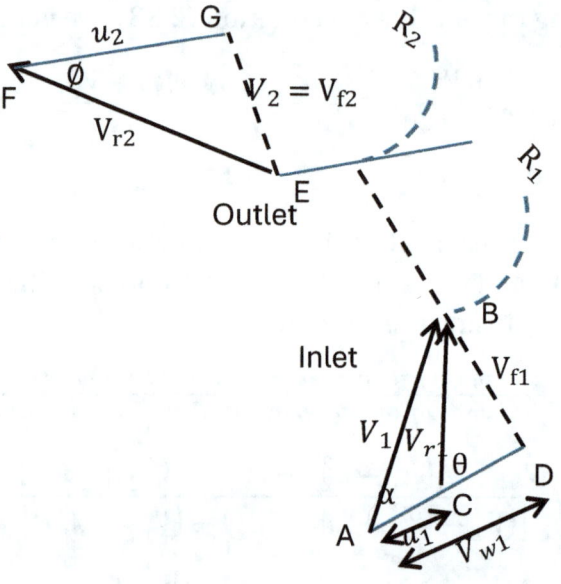

Figure 13: Velocity Triangles at inlet and outlet.

There is not much change in velocity of flow,

Therefore $V_{f1} = V_{f2}$

$$H_e = \left(\frac{1}{g}\right)V_{w1}.u_1$$

$$= \frac{1}{g}[V_{f1}.Cot\alpha][V_{f1}Cot\alpha - V_{f1}Cot\theta]$$

Therefore, $H_e = \frac{1}{g}V_{f1}^2 Cot\alpha[Cot\alpha - Cot\theta]$

Also, $(V_1^2 - V_2^2) = (V_{f1} Cosec\alpha)^2 - V_{f2}^2$; Because $V_2 = V_{f2}$

$$= V_{f1}^2[Cosec^2\alpha - 1]; \text{ Because } V_{f1} = V_{f2}$$

$$= V_{f1}^2 Cot^2\alpha$$

$$R = 1 - [V_{f1}^2[Cosec^2\alpha - 1]/[2g\left[\frac{1}{g}V_{f1}^2 Cot\alpha[Cot\alpha - Cot\theta]\right]$$

$$R = 1 - \frac{[Cot\alpha\}}{[2(Cot\alpha - Cot\theta)]}$$

2. 38

Definitions:

(i)   Speed ratio = $(u_1)/[(2gH)^{0.5}]$
(ii)   Flow ratio = $(V_{f1})/[(2gH)^{0.5}]$
(iii)   Discharge of Turbine Q $=\pi D_1 B_1.V_{f1} = \pi D_2 B_2.V_{f2}$
        Where B is the width of runner.
(iv)   If thickness of vanes is taken into consideration.

        Then discharge of turbine = $(\pi D - nt)B_1.V_{f1}$;

        N is the number of vanes and t is the thickness of each vane.

(v)   The head (H) on the turbine is given by

$$H = \frac{p_1}{\rho g} + \frac{V_1^2}{2g}$$

2.39

Axial Flow Reaction Turbine:

If the water flows parallel to the axis of the rotation of the shaft, the turbine is known as axial flow turbine. And if the head at the inlet of the turbine is the sum of pressure energy and kinetic energy and during the flow of water through runner, a part of pressure energy is converted into kinetic energy, the turbine is known as a reaction turbine.

For the axial flow reaction turbine, the shaft of the turbine is vertical. The lower end of the shaft is made longer, which is known as hub or boss.

The vanes are fixed on the hub and hence hub acts as a runner for axial flow reaction turbine. The following are the important type of axial flow reaction turbines:

1. Propeller Turbine and
2. Kaplan Turbine.

   When the vanes are fixed to the hub and they are not adjustable, the turbine is known as Propeller turbine.

   But if the vanes on the hub are adjustable, the turbine is known as Kaplan turbine after the name of V. Kaplan, an Austrian engineer.

   This turbine is suitable where a large quantity of water at low head is available.

   Figure 14 shows the runner of a Kaplan turbine, which consists of a hub fixed to the shaft. On the hub, the adjustable vanes are fixed as shown in figure.

   The main parts of a Kaplan turbine are:
   1. Scroll casing.
   2. Guide vanes mechanisms.
   3. Hub with vanes or runner of the turbine and
   4. Draft Tube.

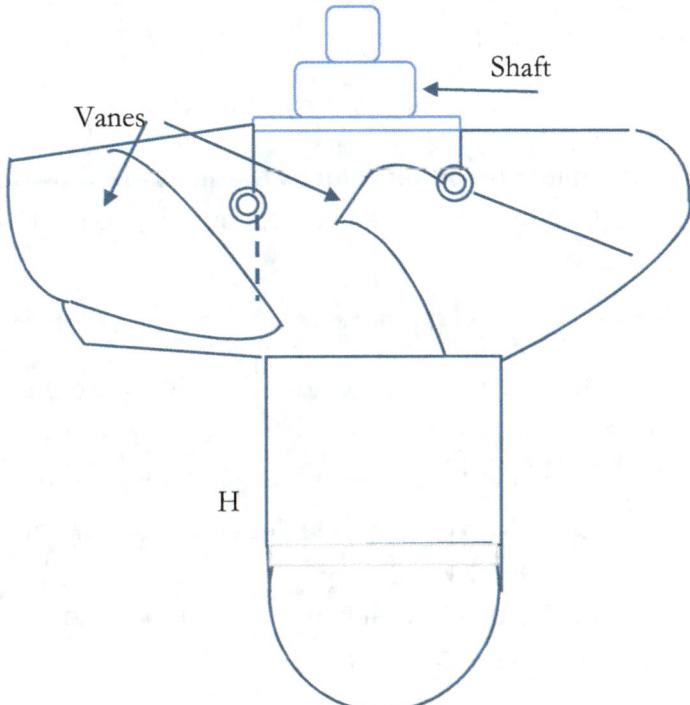

Figure 14: Kaplan Turbine Runner.

Figure 14 shows all the main parts of a Kaplan turbine. The water from penstock enters the scroll casing and then moves to the guide vanes. From the guide vanes, the water turns through 90° and flows axially through the runner as shown in figure.

The discharge through the runner is obtained as

$$Q = \frac{\pi}{4}(D_o^2 - D_b^2).V_{f1}$$

Where, D_o = Outer diameter of runner.

D_b = Diameter of hub.

V_{f1} = Velocity of flow at inlet.

The inlet and outlet velocity triangles are drawn at the extreme edge of the runner vane corresponding to the points 1 and 2 as shown in figure.

Some important points:

1. The peripheral velocity at inlet and outlet are equal;

$$u_1 = u_2 = \frac{\pi D_o N}{60}$$

2. Velocity of flow at inlet and outlet are equal
   That is $V_{f1} = V_{f2}$
3. Area of flow at inlet and outlet are equal.
   That is $\frac{\pi}{4}(D_o^2 - D_b^2)$

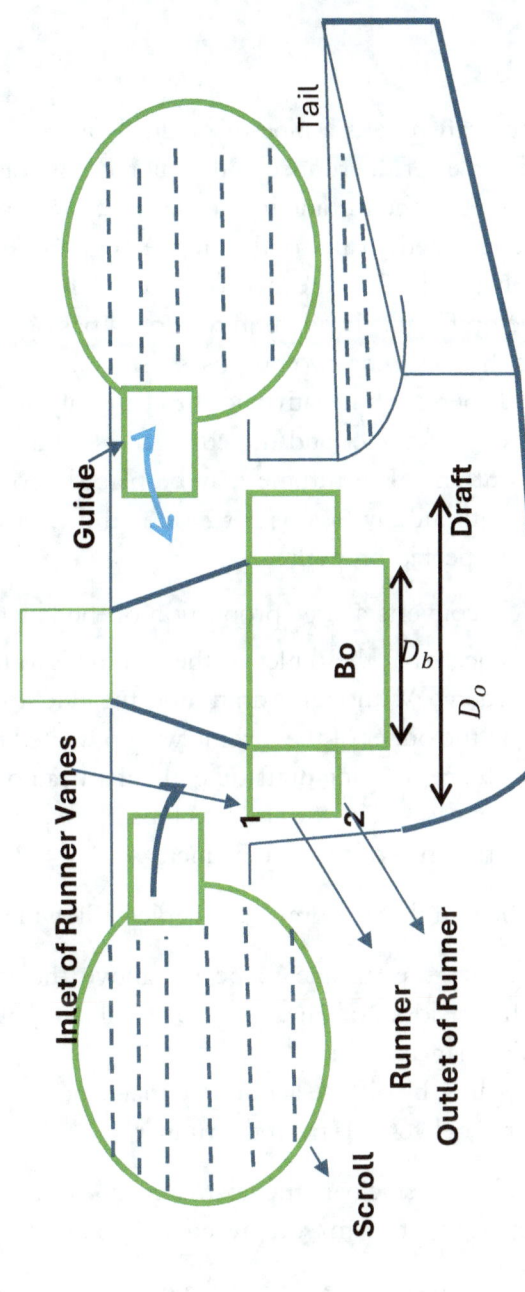

Figure 15: Main Components of a Kaplan Turbine.

Draft Tube:

The draft tube is a pipe of gradually increasing area. It connects the outlet of the runner to the tail race and discharges water from the exit of the turbine to the tail race. One end of the draft- tube is connected to the outlet of the runner while the other end is submerged below the level of water in the tail race.

The draft-tube, in addition to serve a passage for water discharge, has the following two purposes also.

1. It permits a negative head to be established at the outlet of the runner and thereby increase the net head on the turbine. The turbine may be placed above the tail race without any loss of net head and hence turbine may be inspected properly.

2. It converts a large proportion of the kinetic energy $\left(\frac{V_2^2}{2g}\right)$ rejected at the outlet of the turbine into useful pressure energy. Without the draft tube, the kinetic energy rejected at the outlet of the turbine will go wasted to the tail race. Hence by using draft-tube, the net head on the turbine increases. The turbine develops more power and also the efficiency of the turbine increases.

It can also be said that the draft tube serves following two purposes:

1. It allows the turbine to be set above the tail-water level, without loss of head, to facilitate inspection and maintenance.

2. It regains, by diffuse action, the major portion of the kinetic energy delivered it from the runner.

At rated load, the velocity at the upstream end of the tube for modern units ranges from 7 to 9 m/s, representing from 2.7 to 4.8m head.

Good practice limits the velocity at the discharge end of the tube from 1.5 to 2.1 m/s, representing less than 0.3 m velocity head loss.

Types of Draft Tubes:

1. Conical Draft-tube.

2. Simple Elbow Tube.

3. Moody Spreading Tube and

4. Draft-tube with Circular Inlet and Rectangular Outlet.

Conical draft tubes and Moody spreading draft tubes are most efficient while simple elbow tubes and elbow draft tubes with circular inlet and rectangular outlet require less space compared to other draft tubes.

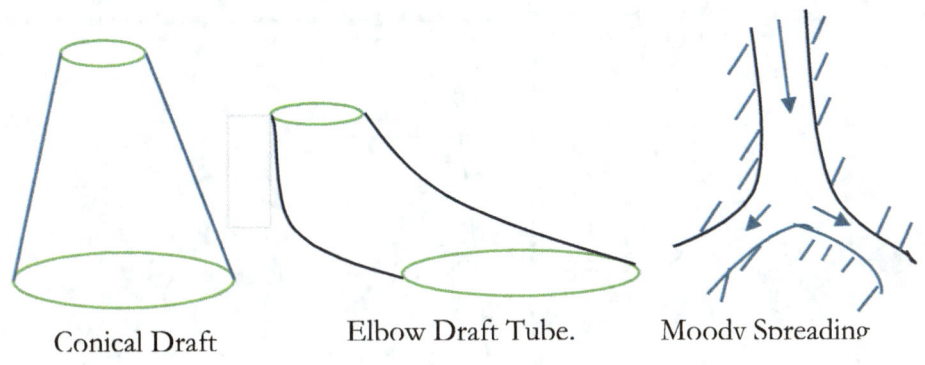

Conical Draft   Elbow Draft Tube.   Moody Spreading

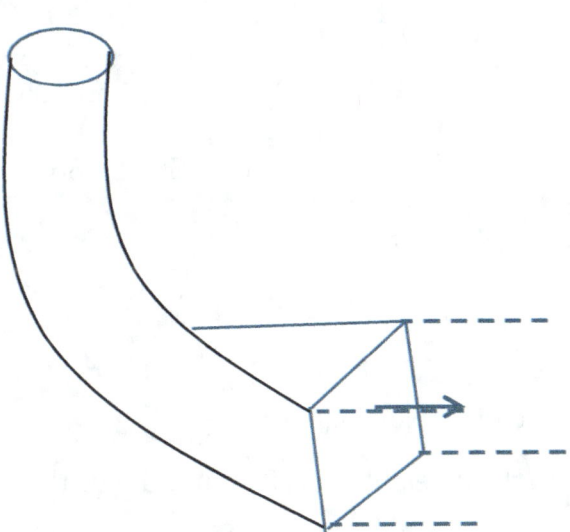

Draft Tube with Circular Inlet and Rectangular Output.

Draft-tube Theory:

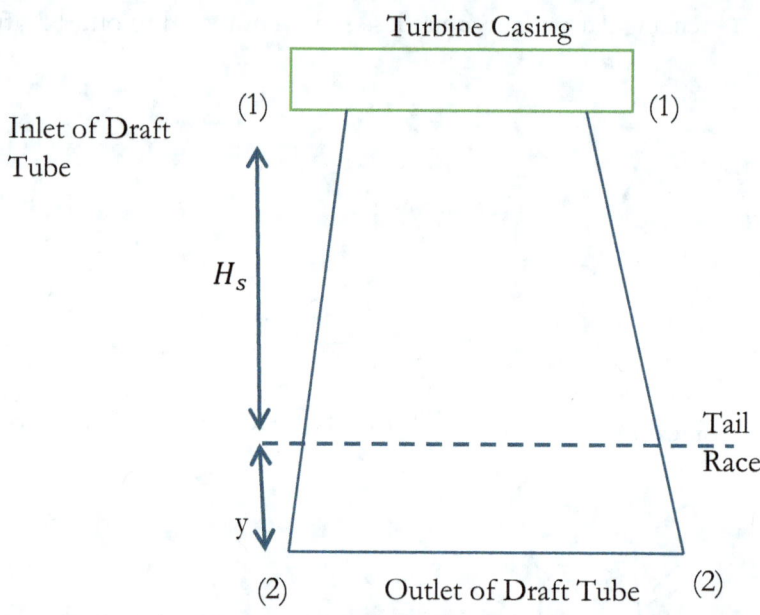

Figure 17: Draft Tube Theory

Figure 17 shows a typical draft tube.

Let H_s = Vertical height of draft — tube above the tail race.

y = Distance of bottom of draft — tube from tail — race.

Applying Bernoulli's equation to inlet (section 1-1) and outlet (section 2-2) of draft-tube and taking section 2-2 as datum line. We get,

$$\frac{p_1}{\rho g} + \frac{V_1^2}{2g} + (H_s + y) = \frac{p_2}{\rho g} + \frac{V_2^2}{2g} + 0 + h_f$$

2.40

Where, h_f = loss of energy between sections 1-1 and 2-2.

But $\frac{p_2}{\rho g}$ = Atmospheric pressure + y = $\frac{p_a}{\rho g}$ +y

Therefore,

$$\frac{p_1}{\rho g} + \frac{V_1^2}{2g} + (H_s + y) = \frac{p_a}{\rho g} + y + \frac{V_2^2}{2g} + h_f$$

Therefore,

$$\frac{p_1}{\rho g} = \frac{p_a}{\rho g} - H_s - [\frac{V_1^2}{2g} - \frac{V_2^2}{2g} - h_f]$$

2.41

Therefore $\frac{p_1}{\rho g}$ is less than atmospheric pressure, resulting in diffusion.

Efficiency of Draft-tube:

The efficiency of a draft-tube is defined as the ratio of actual conversion of kinetic energy head into pressure energy head in the draft-tube to the kinetic energy head at the inlet of the draft-tube.

Mathematically

$$\eta_h = \frac{Actual\ conversion\ of\ kinetic\ head\ into\ pressure\ head}{Kinetic\ head\ at\ the\ inlet\ of\ draft - tube}$$

Theoretical conversion of kinetic head into pressure head in draft-tube $= \frac{V_1^2}{2g} - \frac{V_2^2}{2g}$

61

Actual conversion of kinetic head into pressure head

$$= \frac{V_1^2}{2g} - \frac{V_2^2}{2g} - h_f$$

$$\eta_h = \frac{\left[= \frac{V_1^2}{2g} - \frac{V_2^2}{2g} - h_f \right]}{\frac{V_1^2}{2g}}$$

2.42

Specific Speed:

The specific speed is used in comparing the different types of turbines.

It is defined as the speed of a turbine, which is identical in shape, geometrical dimensions, blade angles, gate opening etc. with the actual turbine but of such a size that it will develop unit power when working under unit head. It is denoted by the symbol N_s.

In MKS units, unit power is taken as one horsepower and unit head as one meter. But in S.I. units, unit power is taken as one kilowatt and unit head as one meter.

Derivation of Specific Speed:

The overall efficiency of a turbine is given by

$$\eta_o = \frac{Shaft\ Power}{Water\ Power} = \frac{\dfrac{Power\ Developed}{\rho g Q H}}{1000}$$

Where,

H is the head under which turbine is working

Q is the discharge through turbine.

P is the power developed or shaft power.

Therefore, $P = \eta_o \cdot \dfrac{\rho g Q H}{1000}$ or $P \propto QH$

2.43

Now denoting as following,

D is the diameter of actual turbine.

N is the speed of actual turbine.

u is the tangential velocity of turbine.

V is the absolute velocity of water.

N_s is the specific speed.

The absolute velocity, tangential velocity and head on turbine are related as,

$$u \propto V \text{ and } V \propto \sqrt{H}; also \ u \propto \sqrt{H}$$

2.44

$$u = \frac{\pi DN}{60} \propto DN$$

2.45

From equations (2.44 & 2.45), we have

$$\sqrt{H} \propto DN \text{ or } D \propto \frac{\sqrt{H}}{N}$$

2.46

The discharge through turbine is given by

Q = Area x Velocity

Area $\propto BxD \propto D^2$

Where B is the width of runner and also $B \propto D$

And $Velocity \propto \sqrt{H}$

Therefore $Q \propto D^2\sqrt{H}$

From equation (2.45) $Q \propto \dfrac{H\sqrt{H}}{N^2} \propto \dfrac{H^{\frac{3}{2}}}{N^2}$

63

From equation (2.43).

$$\frac{P}{H} \propto \frac{H^{\frac{3}{2}}}{N^2}$$

$$P = k\frac{H^{\frac{5}{2}}}{N^2}$$

Where k is the constant of proportionality.

From definition of specific speed if P = 1, H =1, the speed N = Specific speed N_s.

Therefore, $1 = k.\frac{1}{N_s^2}$

Or $N_s^2 = k$

Therefore $P = N_s^2\left[\frac{H^{\frac{5}{2}}}{N^2}\right]$

Or $N_s^2 = \frac{N^2P}{H^{\frac{5}{2}}}$

Therefore, $N_s = \frac{N\sqrt{P}}{H^{\frac{5}{4}}}$

2.47

If P is taken in metric horsepower, the specific speed is obtained in MKS units and if P is taken in Kilowatts, the specific speed is obtained in S.I. units.

Unit Quantities:

In order to predict the behaviour of a turbine working under varying conditions of head, speed, output and gate opening, the results are expressed in terms of quantities which may be obtained when head on the turbine is reduced to unity. The conditions of turbine under unit head are such that the efficiency of the turbine remains unaffected. The following are the three important unit quantities which must be studied under unit head:

64

1. Unit speed.
2. Unit discharge and
3. Unit power.

Unit Speed:

It is defined as the speed of a turbine working unit head (i.e. under a head of 1m). It is denoted by N_u.

Let,

N = Speed of a turbine under a head H.

H= Head under which a turbine is working.

u= Tangential velocity of turbine.

The tangential velocity, absolute velocity of water and head on turbine are related as follows,

$$u \propto V$$

$$V \propto \sqrt{H}$$

Therefore $u \propto \sqrt{H}$

Also, tangential velocity $u = \frac{\pi DN}{60}$

Therefore $u \propto N$ or $N \propto u \propto \sqrt{H}$

Therefore $N = k_1.\sqrt{H}$

2.48

Where k_1 is constant of proportionality.

If head on turbine becomes unity, the speed N becomes unit speed N_u.

That is H = 1 and N = N_u

So, from equation (2.48)

$$N_u = k_1\sqrt{1}$$

Therefore, $N = N_u\sqrt{H}$

Or $N_u = \dfrac{N}{\sqrt{H}}$

$$2.49$$

Unit Discharge:

It is defined as the discharge passing through a turbine, which is working under a unit head (i.e. 1m). It is denoted by the symbol Q_u.

Let H = head of water on turbine.

Q = Discharge passing through turbine, when head is H on turbine.

a = Area of flow of water.

The discharge passing through a given turbine, under a head 'H' is given by,

Q = Area of flow x Velocity.

But for a turbine, area of flow is constant, and velocity of flow is proportional to $\sqrt{H}$.

Therefore, $Q \propto Velocity \propto \sqrt{H}$

Therefore, $Q = k_2\sqrt{H}$

Where k_2 is constant of proportionality.

If $H = 1, Q = Q_u$

$$Therefore\ Q_u = k_2$$

$$Or\ Q_u = \dfrac{Q}{\sqrt{H}}$$

$$2.50$$

Unit Power:

It is defined as the power developed by a turbine, working under a unit head (i.e. 1m).

Let P = Power developed by the turbine under a head H and discharge Q.

From equation (2.43) of overall efficiency relation, we get

$$P = \eta_o \cdot \frac{\rho g Q H}{1000}$$

Therefore $P \propto Q.H \propto \sqrt{H}.H$, because $Q \propto H$

$$\text{Therefore } P \propto H^{3/2}$$

Therefore, $P = k_3 H^{3/2}$, k_3 is the Constant of proportionality.

When H =1, $P = P_u$

$$\text{Thus } P_u = k_3$$

Therefore $P = P_u H^{3/2}$

$$Or, P_u = \frac{P}{H^{\frac{3}{2}}}$$

2.51

Use of Unit Quantities ($N_u Q_u P_u$):

If a turbine is working under different heads, the behaviour of the turbine can be easily known from the values of unit speed, unit discharge and unit power.

H_1, H_2 are the heads under which turbine works.

N_1, N_2 are corresponding speeds.

Q_1, Q_2 are corresponding discharges.

P_1, P_2 are the corresponding power generated.

$$N_u = \frac{N_1}{\sqrt{H_1}} = \frac{N_2}{\sqrt{H_2}}$$

$$Q_u = \frac{Q_1}{\sqrt{H_1}} = \frac{Q_2}{\sqrt{H_2}}$$

$$P_u = \frac{P_1}{H_1^{\frac{3}{2}}} = \frac{P_2}{H_2^{\frac{3}{2}}}$$

2.52

Hence, if the speed, discharge and power developed by turbine under a head are known, then by equations (2.52), the speed, discharge and power developed by the same turbine under a different head can be obtained easily.

Characteristic Curves of Hydraulic Turbines:

Characteristic curves of a hydraulic turbine are the curves, with the help of which the exact behaviour and performance of the turbine under different working conditions can be known.

These curves are plotted from the results of tests performed on the turbine under different working conditions.

The important parameters which are varied during a test on turbine are:

1. Speed (N) 2. Head (H) 3. Discharge (Q) 4. Power (P) 5. Overall efficiency (η_o) 6. Gate opening.

Out of these six parameters, three parameters namely Speed (N), Head (H) and Discharge (Q) are independent parameters. Out of the three independent parameter (N, H & Q), one parameter is kept constant (say H) and the variation of other four parameters with respect to any one of the remaining two independent variables (say N and Q) are plotted and various curves are obtained. These curves are called characteristic curves.

The following are the important characteristic curves of a turbine.

1. Main characteristic curves or Constant head curves.
2. Operating characteristic curves or Constant speed curves.
3. Muschel curves or Constant efficiency curves.

Main Characteristic Curves (H = Constant):

Main characteristic curves are obtained by maintaining a constant head and a constant gate opening (G.O.) on the turbine. The speed of turbine is varied by changing load on the turbine.

For each value of the speed, the corresponding values of the power (P) and discharge (Q) are obtained. Then overall efficiency (η_o) for each value of speed is calculated.

From these readings the values of unit speed (N_u), unit power (P_u) and unit discharge (Q_u) are determined.

Taking N_u as abscissa, the values of Q_u, P_u, P and η_o are plotted as shown in figures....

By changing the gate opening, the values of Q_u, P_u, P and η_o are obtained and taking N_u as abscissa, these values are plotted.

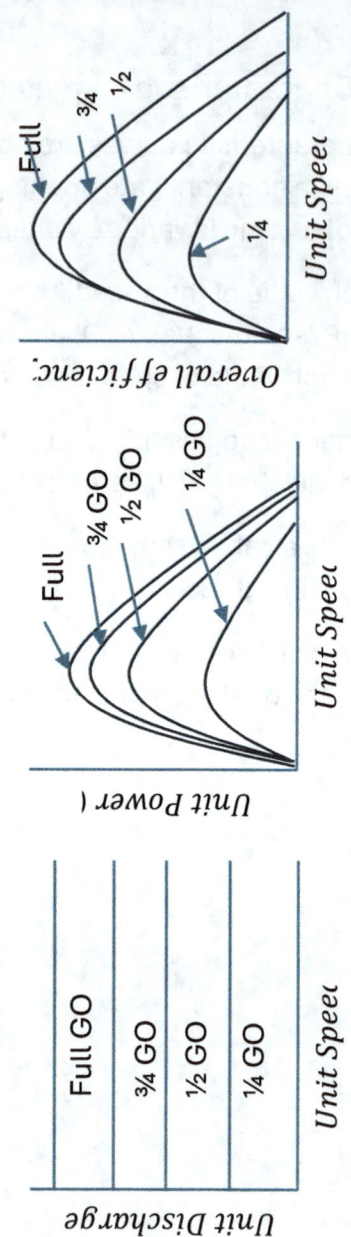

Figure 18: Main Characteristic Curves for a Pelton Turbine (H = Constant).

70

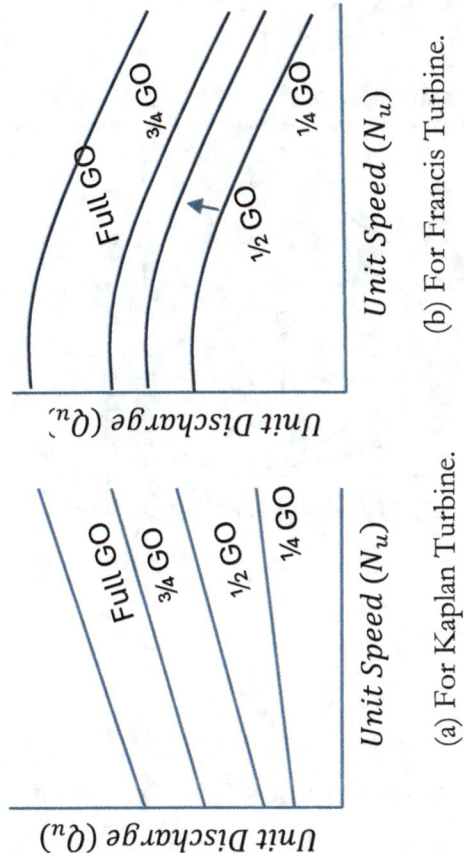

(a) For Kaplan Turbine.   (b) For Francis Turbine.

Figure19: Main Characteristics Curves for Reaction (Kaplan & Francis) Turbines (H=Constant).

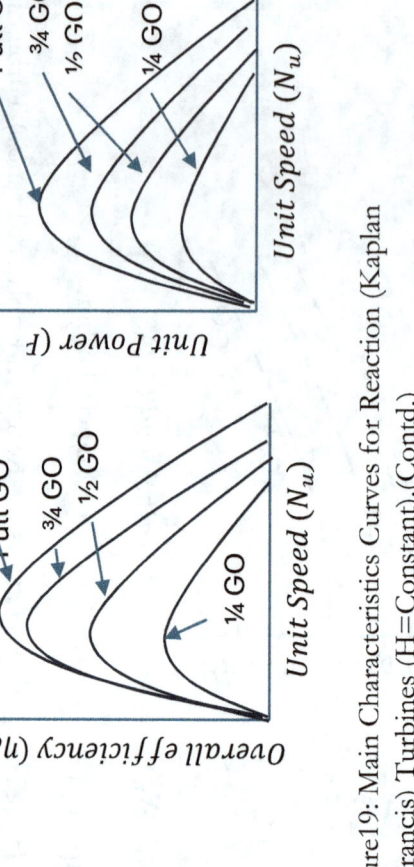

Figure19: Main Characteristics Curves for Reaction (Kaplan & Francis) Turbines (H=Constant).(Contd.)

Operating Characteristics (N = Constant):

In case of turbines, generally head is constant. For operating characteristics N and H are constant and hence variation of power and efficiency with respect to discharge Q are plotted.

The power curve for turbines shall not pass through the origin because certain amount of discharge is needed to produce power to overcome initial friction. Hence power and efficiency curves will be slightly away from the origin on the x-axis, as to overcome initial friction, certain amount of discharge will be required.

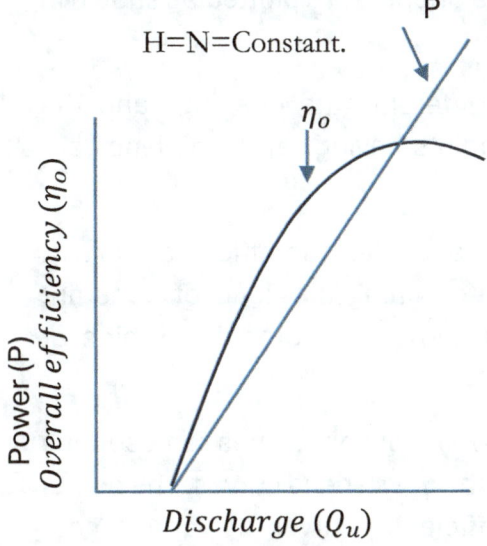

H=N=Constant.

P

η_o

Power (P)
Overall efficiency (η_o)

Discharge (Q_u)

Figure 20: Operating Characteristic Curves.

Constant Efficiency or Muschel Curves or Iso-efficiency Curves:

These curves are obtained from the speed vs efficiency and speed vs discharge for different gate openings.

For a given efficiency from the N_u vs η_o curves, there are two speeds. From N_u vs Q_u curves, corresponding to two values of speed, there are two values of discharge. Hence for a given efficiency there are two values of discharge for a particular gate opening.

This means that for a given efficiency there are two values of efficiency and two values of discharge for a given gate opening. If the efficiency is maximum, there is only one value.

73

These two values of speed and two values of discharge corresponding to a particular gate opening are plotted as shown in figure.

The procedure is repeated for different gate openings and the curves Q vs N are plotted. The points having same efficiency are joined.

The curves having same efficiency are called Iso-efficiency curves. These curves are helpful for determining the zone of constant efficiency and for predicting the performance of the turbine at various efficiencies.

For plotting the iso-efficiency curves, horizontal lines representing the same efficiency are drawn on the η_o vs speed curves. The points at which these lines cut the efficiency curves at various gate openings are transferred to the corresponding Q vs speed curves. The points having the same efficiency are then joined by a smooth curve. These smooth curves represent the Iso-efficiency curve.

Pelton Turbine:

The discharge Q_u depends only upon the gate opening and is independent of N_u, the curves for Q_u are horizontal.

Francis Turbine:

The curves between Q_u and N_u for Francis turbine are falling curves. This is due to fact that a centrifugal head develops, which acts outward and opposes the external head causing flow eventually decreasing the discharge as the speed increases.

The curves between P_u & N_u and these between η_o & N_u indicate that at a particular speed the efficiency is maximum.

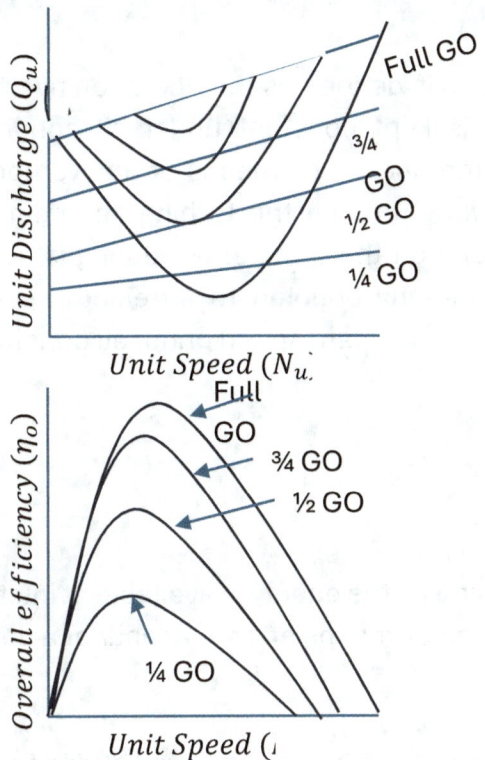

Figure 21: Constant Efficiency Curve.

Kaplan Turbine:

The curves between Q_u and N_u for Kaplan turbine are rising curves, the discharge increases with the increase in speed.

The curves between P_u & N_u and these between η_o & N_u indicate that at a particular speed the efficiency is maximum.

Governing of Turbines:

The governing of turbines is defined as the operation by which the speed of the turbine is kept constant under all conditions of working. It is done automatically by means of a governor, which regulates the rate of flow through the turbine according to the changing load conditions on the turbine. Governing is necessary because a turbine is directly coupled to an electric generator, which is required to run at constant speed under all conditions.

Impulse Turbine:

In case of impulse turbine all the energy is available at inlet, so flow in nozzle is regulated by means of a spear and/ or a deflector.

Reaction Turbine:

In case of a reaction turbine, energy is generated during flow in the runner, so energy is regulated by means of guide vanes.

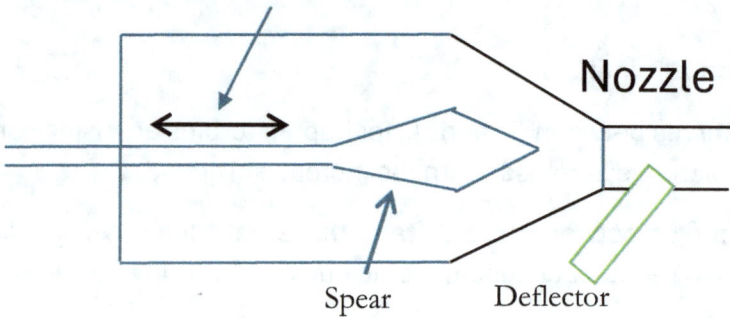

Figure 22: Spear & Deflector Regulation of Impulse Turbine.

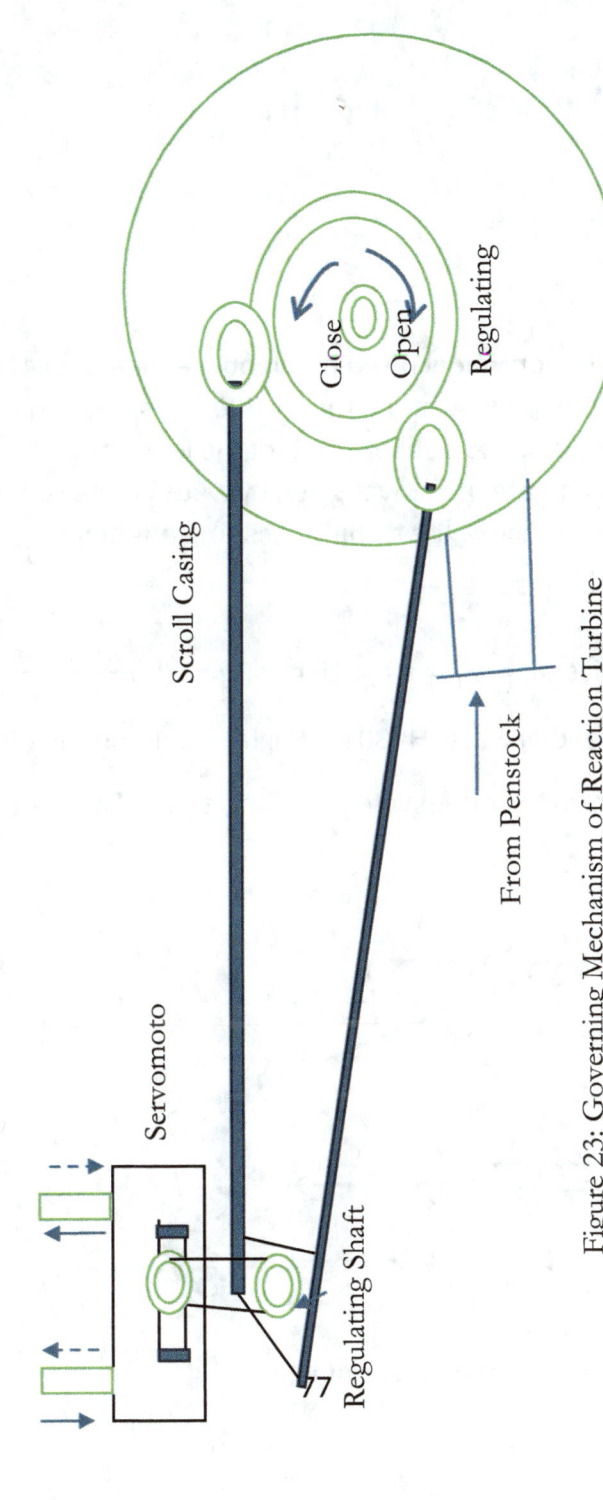

Figure 23: Governing Mechanism of Reaction Turbine

Worked Examples

Problem 7:

A Pelton wheel has a mean bucket speed of 10 m/s with a jet of water flowing at the rate of 700 liters/second under a head of 30 meters. The buckets deflect the jet through an angle of 160°. Calculate the power given by water to the runner and the hydraulic efficiency of the turbine. Assume coefficient of velocity as 0.98.

Solution:

Given $u = u_1 = u_2 = 10$ m/s; $Q = 700\frac{litres}{s} = \frac{0.7m^3}{s}$; $C_v = 0.98$.

Head of water H=30 m, Angle of deflection 160°.

From the outlet velocity triangle, $\emptyset = 180° - 160° = 20°$.

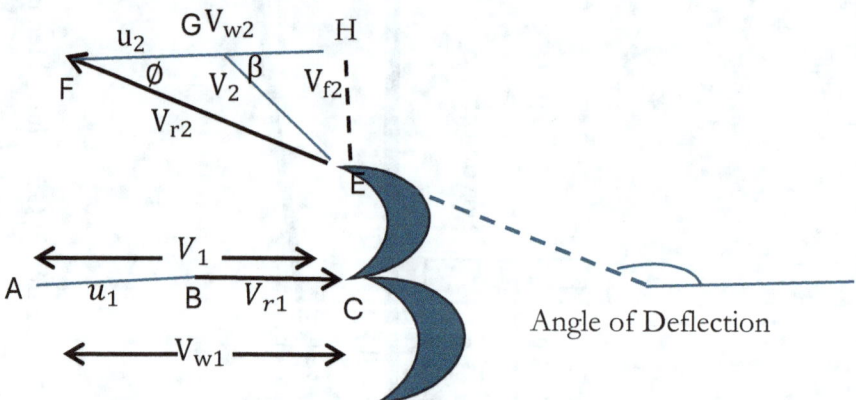

Figure: Problem 7

From inlet velocity triangle,

$$V_1 = C_v\sqrt{2gH} = 0.98\sqrt{2x9.81x30} = 23.77\frac{m}{s}.$$

Therefore, $V_{r1} = V_1 - u_1 = 23.77 - 10 = 13.77\frac{m}{s}.$

From the outlet velocity triangle,

$$V_{r1} = V_{r2} = 13.77\frac{m}{s}.$$

$$V_{w2} = V_{r1}Cos\emptyset - u_2 = 13.77Cos20^o - 10.0 = 2.94\frac{m}{s}.$$

Work done by the jet per second on the runner=$\rho a V_1(V_{w1} + V_{w2})$

$$.u$$

$$= 1000x0.7x[23.77 + 2.94]x10 \qquad (where, aV_1 = Q)$$

=186970 Nm/s

Power given to turbine= Work done/1000=186.97 Kw

The hydraulic efficiency, $\eta_h = \frac{[2(V_{w1}+V_{w2})xu]}{V_1^2} =$
$[2(23.77 + 2.94)x10]/(23.77)^2$

$$=0.9454 \text{ or } 94.54\%$$

Problem 8:

A Pelton wheel is to be designed for the following specifications:

Shaft power= 11,772 Kw, Head = 380 m, Speed= 750 rpm, Overall efficiency= 86%, Jet diameter is not to exceed One-sixth of wheel diameter.

Determine: (i) The wheel diameter, (ii) Number of jets required and (iii) the diameter of jet. Take $k_v = 0.98$ and $k_{u1} = 0.45$

Solution:

Shaft Power = 11,772 kw, Head H=380 m, Speed N= 750 rpm, η_o = 0.86 and ratio of jet diameter to wheel diameter d/D=1/6.

Velocity of jet, $V_1 = C_v.\sqrt{2gH} = 0.985\sqrt{2x9.81x380} = 85.05$ m/s

Tangential velocity $u = k_{u1}\sqrt{2gH} = 38.85 \ m/s$

Also $u = \dfrac{\pi DN}{60}$

Therefore $D = \dfrac{60.u}{N} = 0.989 \ m$

Also, diameter of jet $d = \dfrac{D}{6} = 0.165 \ m$

Discharge of one jet $q = \dfrac{\pi d^2 V_1}{4} = 1.818 \ m^3/s$

Now $\eta_o = \dfrac{Shaft\ Power}{Water\ Power} = \dfrac{\frac{11772 \times 1000}{\rho g Q H}}{1000} = 0.86$

Therefore, total discharge $Q = 3.672 m^3/second$

Number of jets $= \dfrac{Total\ discharge}{Discharge\ from\ one\ jet} = \dfrac{3.672}{1.818} = 2 \ Jets.$

Problem 9:

The penstock supplies water from a reservoir to the Pelton Wheel with a gross head of 500 m. One third of the gross head is lost in friction in the pen stock. The rate of flow of water through the nozzle fitted at the end of penstock is 2.0 m^3/s. The angle of deflection of the jet is 165°. Determine the power given by the water to the runner and also hydraulic efficiency of the Pelton wheel. Take speed ratio =0.45 and $C_v = 1.0$.

Solution:

$$H_g = 500 \ m; \ h_f = \dfrac{H_g}{3} = 166.7 \ m; Q = \dfrac{2.0m^3}{s}; \emptyset = 180^o - 165^o$$
$$= 15^o$$

Net head $= H_g - h_f = 333.30 \ m$

Velocity of jet $V = C_v\sqrt{2gH} = 1.0\sqrt{2 \times 9.81 \times 333.3} = 80.86 \ m/s$

Velocity of wheel $u = speed\ ratio\sqrt{2gH} = 0.45\sqrt{2 \times 9.81 \times 333.3}$
$$= 36.387 \ m/s$$

$V_{r1} = V_1 - u_1 = 80.86 - 36.387 = 44.473 \ m/s$

$V_{w1} = V_1 = 80.8 \ m/s$

From outlet velocity triangle:

$V_{r2} = V_{r1} = 44.473 \ m/s$
$V_{r2}Cos\emptyset = u_2 + V_{w2}$
$44.473Cos15^o = 36.387 + V_{w2}$
$Therefore, V_{w2} = 6.57 \ m/s$

Work done by the jet on the runner per second $= \rho aV_1[V_{w1} + V_{w2}]$
$u = \rho Q[V_{w1} + V_{w2}]u$

$= 6362630 \ Nm/s$

$$The \ power \ given \ the \ water \ to \ the \ runner = \frac{Work \ done}{1000}$$

$= 6362.63 \ kw$

Hydraulic efficiency of the turbine $\eta_h = \frac{[2(V_{w1}+V_{w2}).u]}{V_1^2}$

$= 0.9731 \ or \ 97.31\%$

Problem 10:

A Pelton wheel has a mean bucket diameter of 1 m and is running at 1000 rpm. The net head on the Pelton wheel is 700 m. If the side clearance angle is 15° and discharge through nozzle is 0.1 m^3/s. Find the Power available at the nozzle and hydraulic efficiency of the turbine.

Solution:

Diameter of wheel D= 1.0 m; Speed N = 1000 rpm; Net head = 700 m; $\emptyset = 15^o; Q = 0.1\frac{m^3}{s}$.

Tangential velocity $u = \frac{\pi DN}{60} = 52.36 \ m/s$

Velocity of jet at inlet $V_1 = C_v\sqrt{2gH} = 117.19 \ m/s$

Power available at nozzle $= \frac{WH}{1000} = \frac{\rho gQH}{1000} = 686.7 \ kw$

Hydraulic efficiency of the turbine $\eta_h = \frac{[2(V_{w1}+V_{w2}).u]}{V_1^2}$

$V_{w1} = V_1$
$V_{w2} = V_{r1}Cos\emptyset - u = (V_1 - u)Cos\emptyset - u$
Therefore,

Hydraulic efficiency of the turbine $\eta_h = \dfrac{[2(V_{w1}+V_{w2}).u]}{V_1^2} =$

$$\dfrac{[2(V_1+((V_1-u)Cos\emptyset-u).u]}{V_1^2}$$

$$= \dfrac{[2(V_1-u)(1+Cos\emptyset).u]}{V_1^2} = 0.9718 \ or \ 97.18\%$$

Problem 11:

A Pelton wheel is working under a gross head of 400 m. The water is supplied through penstock of diameter 1 m and length 4 km from reservoir to the Pelton wheel. The coefficient of friction for the penstock is given as 0.008. The jet of water diameter 150 mm strikes the buckets of the wheel and gets deflected by 165°. The relative velocity of water at outlet is reduced by 15% due to friction between inside surface of the bucket and water. If the velocity of the buckets is 0.45 times the jet velocity at inlet and mechanical efficiency as 85%, determine (i) Power given to the runner, (ii) Shaft Power, (iii) Hydraulic efficiency & Overall efficiency.

Solution:

$\emptyset = 180^o - 165^o = 15^o$

$V_{r2} = 0.85V_{r1}; u = 0.45 x Jet \ velocity.$

$Let \ V^* = velocity \ of \ water \ in \ Penstock.$

$Using \ continuity \ equation; Area \ of \ penstock x V^*$

$$= Area \ of \ jet \ x \ V_1$$

$Therefore, V^* = \dfrac{d^2}{D^2} x V_1 = 0.0225 V_1$ [where, D is the diameter of Penstock.]

Applying Bernoulli's equation to the free surface of water in reservoir and outlet nozzle.

$H_g = Head \ lost \ due \ to \ friction + \dfrac{V_1^2}{2g}$

$400 = (4 fl 400 = \dfrac{4 f l V^{*2}}{2gD} + \dfrac{V_1^2}{2g} = \dfrac{4 x.008 x 4000 x V^*}{1.0 x 2 x 9.81} + \dfrac{V_1^2}{2g} = 0.0543 V_1^2$

Therefore, $V_1 = 85.83$, as $V^* = 0.0225V_1$

Velocity of bucket $u = 0.45V^* = \frac{d^2}{D^2}xV_1 = 0.0225V_1 = 38.62 \, m/s$

From inlet velocity triangle:

$V_{r1} = V_1 - u = 47.21\frac{m}{s}; V_{w1} = V_1 = 85.83\frac{m}{s}.$

From outlet velocity triangle:

$V_{r2} = 0.85V_{r1} = 40.13\frac{m}{s}; V_{w2} = V_{r2}Cos\emptyset - u = 0.143\frac{m}{s}.$

Discharge through nozzle:

$Q = \frac{\pi}{4}d^2xV_1 = 1.516\frac{m^3}{s}.$

Work done on the wheel per second $= \rho aV_1[V_{w1} + V_{w2}]$

$xu = 5033540\frac{Nm}{s}.$

(i)   Power given to the runner $= \frac{Work\ Done}{1000}$

$kw = 5033.54kw$

(ii)   Mechanical efficiency $\eta_m = \frac{Power\ at\ Shaft}{Power\ given\ to\ the\ runner}$

Therefore, Power at shaft $= \eta_m$

$x\ Power\ given\ to\ the\ runner = 0.85x503354$

$= 4278.5\ kw.$

(iii) $\eta_h = \frac{[2(V_{w1} + V_{w2})u]}{V_1^2} = 0.9014; \eta_o = \eta_h x\eta_m = 0.85x0.9014$

$= 0.7662\ or\ 76.62\%.$

Problem 12:

An inward flow reaction turbine has an external and internal diameter as 1 m and 0.5 m respectively. The velocity of flow through the runner is constant and is equal to 1.5 m/s. Determine the Discharge through runner, and width of the turbine at outlet if the width of turbine at inlet is 200 mm.

Solution:

$D_1 = 1\ m$, $D_2 = 0.5\ m$, $V_{f1} = V_{f2} = 1.5 \dfrac{m}{s}$, $B_1 = 200\ mm = 0.2\ m$.

$Discharge = \pi D_1 B_1 V_{f1} = 0.9425 \dfrac{m^3}{s}$.

Also, $\pi D_1 B_1 V_{f1} = \pi D_2 B_2 V_{f2}$ or $D_1 B_1 = D_2 B_2$

Therefore $B_2 = 0.4\ m$ or $400\ mm$.

Problem 13:

An inward flow reaction turbine has external and internal diameters 1 m and 0.6 m respectively. The hydraulic efficiency of turbine is 90%, when the head on the turbine is 36 m. The velocity of flow at outlet is 2.5 m/s and discharge at outlet is radial. If the vane angle at outlet is 15° and the width of the wheel is 100 mm at inlet and outlet, determine: (i) the guide blade angle, (ii) speed of the turbine, (iii) vane angle of the runner at inlet, (iv) volume flow rate of turbine and (v) power developed.

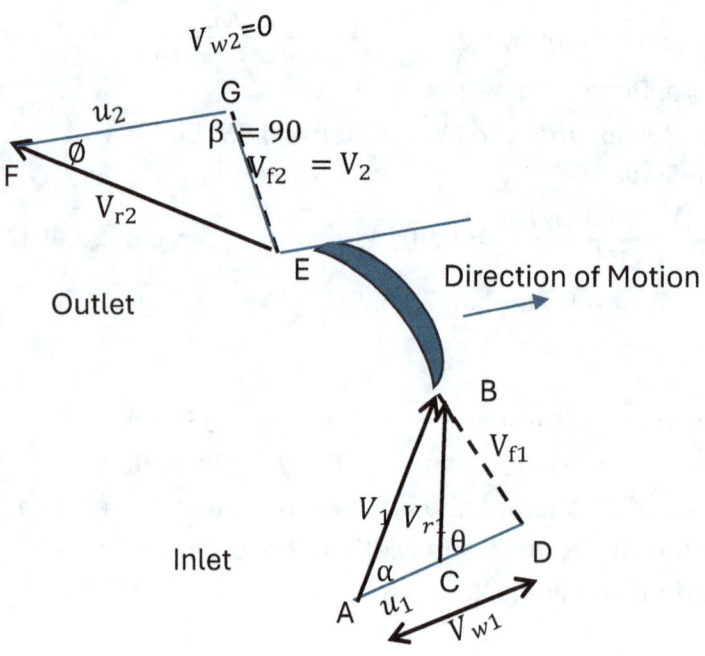

Figure: Problem 13

84

Solution:

$D_1 = 1\ m, D_2 = 0.6\ m, \eta_h = 90\% = 0.9, H = 36\ m, V_{f2}$

$$= 2.5\frac{m}{s}, V_{w2} = 0, \emptyset = 15^o,$$

$B_1 = B_2 = 100\ mm = 0.1\ m$

$\eta_h = \dfrac{V_{w1}u_1}{gH}$ or $0.9 = \dfrac{V_{w1}u_1}{9.81x36}$ or $V_{w1}u_1 = 317.85$

Also, $\tan\emptyset = \dfrac{V_{f2}}{u_2} = \tan15^o = \dfrac{2.5}{u_2}$ or $u_2 = 9.33$

Also, $u_2 = \dfrac{\pi D_2 N}{60}$ or $N = \dfrac{60x9.33}{\pi x0.6} = 296.98\ rpm$

$u_1 = \dfrac{\pi D_1 N}{60} = \dfrac{\pi x1.0x296.98}{60} = 15.55\dfrac{m}{s}$

$V_{w1}u_1 = 317.85, therefore\ V_{w1} = \dfrac{317.85}{15.55} = 20.44\dfrac{m}{s}$

Also, $\pi D_1 B_1 V_{f1} = \pi D_2 B_2 V_{f2}$ or $V_{f1} = \dfrac{D_2 V_{f2}}{D_1} = 1.5\dfrac{m}{s}$

(i)   Guide blade angle α, $Tan\alpha = \dfrac{V_{f1}}{V_{w1}} = \dfrac{1.5}{20.44}$

$= 0.07338$ or $\alpha = Tan^{-1}_{0.07338} = 4.19^o$

(ii)   Speed of turbine N = 296.98 rpm.

(iii)   Angle of runner at inlet$(\theta); \tan\theta = \dfrac{V_{f1}}{V_{w1}-u_1} = \dfrac{1.5}{20.44-15.55}$

$= 0.3067$

(iv)   Volume flow rate $\pi D_1 B_1 V_{f1} = \pi x1x0.1x1.5 = 0.4712$

$\dfrac{m^3}{s}$

(v)   Power developed $= \dfrac{[\rho Q(V_{w1}u_1)]}{1000}kw = 149.76\ kw$

Problem 14:

An inward flow reaction turbine has external and internal diameters as 0.9 m and 0.45 m respectively. The turbine is running at 200 rpm and width of turbine at inlet is 200. The velocity of flow through runner is constant and is equal to 1.8 m/s. The guide blades make an angle of $10°$ to the tangent of the wheel and the discharge at the outlet of the turbine is radial. Draw the inlet and outlet velocity triangles and determine: (i) The absolute velocity of water at inlet of runner. (ii) The velocity of whirl at inlet, (iii) The relative velocity at inlet, (iv) The runner blade angles, (v) width of the runner at outlet, (vi) Mass of water flowing through the runner per second, (vii) head at inlet of turbine, (viii) Power developed and hydraulic efficiency of the turbine.

Solution:

$$D_1 = 0.9 \ m, D_2 = 0.45 \ m, N = 200 \ rpm, B_1 = 200 \ mm = 0.2 \ m, V_{f1}$$
$$= V_{f2} = 1.8 \frac{m}{s}, \alpha = 10°,$$

$Radial \ Discharge, \beta = 90° \ \& \ V_{w2} = 0$

$$u_1 = \frac{\pi D_1 N}{60} = 9.424 \frac{m}{s}; \ u_2 = \frac{\pi D_2 N}{60} = 4.712 \ m/s$$

(i)   Absolute Velocity at inlet:

From inlet velocity triangle, $V_1 Sin\alpha = V_{f1}$ or $V_1 = \frac{18}{Sin10°}$
$= 10.365 \frac{m}{s}.$

(ii)   Velocity of whirl at inlet:

$$V_{w1} = V_1 Cos\alpha = 10.207 \frac{m}{s}.$$

(iii)   Relative velocity at inlet:

$$V_{r1} = \sqrt{V_{f1}^2 + (V_{w1} - u_1)^2} = 1.963 \frac{m}{s}.$$

(iv)   The runner blade angles θ & $\emptyset$.

From inlet velocity triangle, $tan\theta = \frac{V_{f1}}{V_{w1} - u_1}$
$= 2.298 \ or \ \theta = 66° \ 29'$

From outlet velocity triangle, $tan\emptyset = \frac{V_{f2}}{u_2} = \frac{1.8}{4.712} or \ \emptyset =$
$20.9°$

(v)   Width of runner at outlet:

, $\pi D_1 B_1 V_{f1} = \pi D_2 B_2 V_{f2}$ or $B_2 = \frac{D_1 B_1}{D_2}$; because $V_{f1} = V_{f2}$; Therefore $B_2 = 0.4\ m$.

(vi)   Mass flow rate through the runner:

$$Q = \pi D_1 B_1 V_{f1} = \pi x 0.9 x 0.2 x 1.8 = 1.0178 \frac{m^3}{s}.$$

(vii)   Head at inlet of turbine:

$$H - \frac{V_2^2}{2g} = \frac{1}{g}(V_{w1}u_1), Therefore\ H = \frac{V_2^2}{2g} + \frac{1}{g}(V_{w1}u_1)$$

$$= 9.97\ m$$

(viii)   Power developed & Hydraulic efficiency:

$$Power\ developed\ P = \frac{\frac{work\ done}{second}}{1000} = \frac{\rho Q V_{w1}u_1}{1000} = 97.9\ kw$$

$$Hydraulic\ efficiency\ \eta_h = \frac{V_{w1}u_1}{gH} = 0.9834\ or\ 98.34\%.$$

Problem 15:

A Kaplan turbine develops 22000 kw at an average head of 35 m. Assuming a speed ratio of 2, flow ratio of 0.6, diameter of the boss equal to 0.35 times the diameter of the runner and overall efficiency of 88%, Calculate the diameter, speed and specific speed of the turbine.

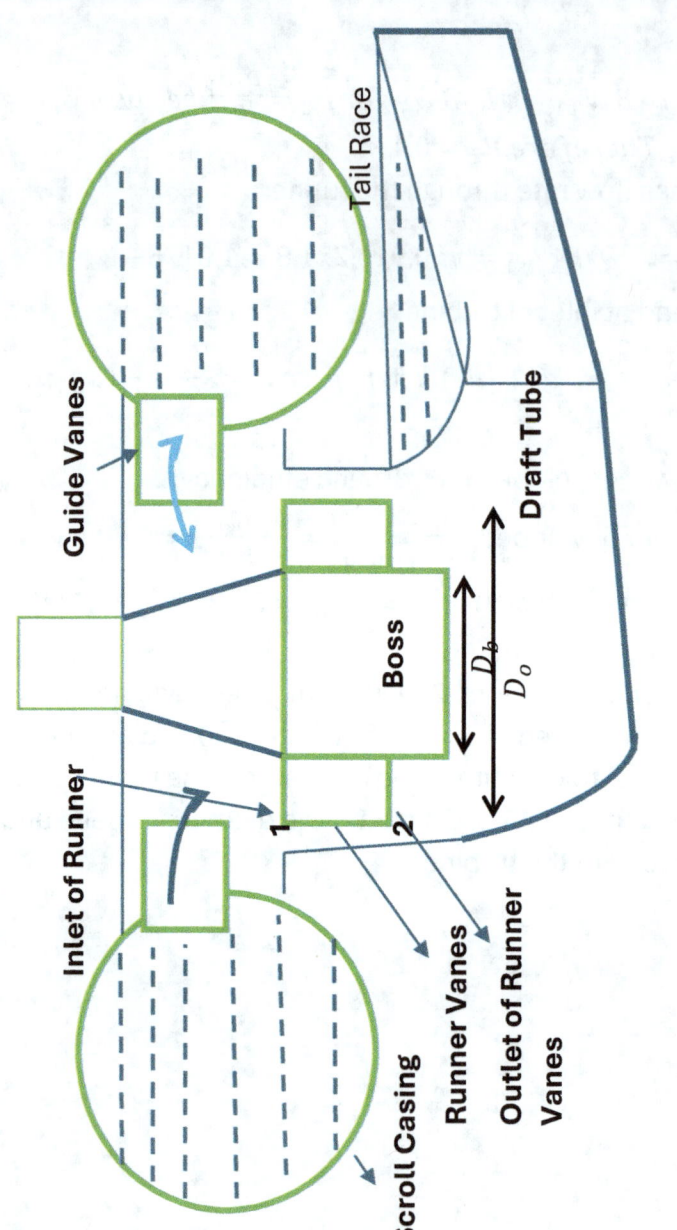

Figure: Problem 15.

Solution:

Shaft power $P = 22000\ kw$, *Head* $H = 35\ m$, *Speed ratio* $k_u = 2.0$, *Flow ratio* $k_f = 0.6$.

Diameter of boss $D_b = 0.35D_o$, *Overall efficiency* η_o
$$= 88\%.$$

$$k_u = \frac{u_1}{\sqrt{2gH}} = 2.0 \text{ or } u_1 = 52.4 \frac{m}{s}$$

$$k_f = \frac{V_{f1}}{\sqrt{2gH}} = 0.6 \text{ or } V_{f1} = 15.7 \frac{m}{s}$$

$$\text{Overall efficiency } \eta_o = \frac{Shaft\ Power}{Water\ Power} = \frac{22000}{\rho g Q H} = 0.88 \text{ or } Q$$

$$= 72.8 \frac{m^3}{s}$$

Also $Q = \frac{\pi}{4}(D_o^2 - D_b^2)V_{f1}$ or $D_o = 2.6\ m$

Speed of turbine; $u_1 = \frac{\pi D_o N}{60} = 52.4$ or $N = 384.9\ rpm$

Specific speed of turbine: $N_s = \frac{N\sqrt{P}}{H^{\frac{5}{4}}} = 670.6$

Problem 16:

Prove that specific speed can be expressed as, $N_s = 3.13N_u$
$\sqrt{Q_u\eta_o}$, where

N_u is unit speed, Q_u is unit discharge and η_o is overall efficiency.
Solution:

$$N_s = \frac{N\sqrt{P}}{H^{\frac{3}{4}}}, \text{ also } P = \rho g Q H x \eta_o = 9.81 Q H \eta_o\ kw$$

Therefore, $N_s = \frac{N}{H^{\frac{3}{4}}}\sqrt{9.81QH\eta_o} = \frac{3.13N}{H^{\frac{5}{4}}}\sqrt{Q\eta_o}$

Also, $N_u = \frac{N}{\sqrt{H}}Q$ and $Q_u = \frac{Q}{\sqrt{H}}$

Also, $N_s = \frac{3.13N}{H^{\frac{5}{4}}}\sqrt{Q\eta_o} = \frac{3.13N}{\sqrt{H}}\sqrt{\frac{Q}{\sqrt{H}}\eta_o} = 3.13N_u\sqrt{Q_u\eta_o}$

Problem 17:

Water leaves the guide vanes of an inward radial flow turbine at an angle α to the tangent to the wheel. The vane angle at entry to the wheel is 90° and velocity of flow at exit is k times that at entry. Prove that, for maximum efficiency under a head H, the peripheral speed should be $\sqrt{\left[\dfrac{2gH}{2+k^2 tan_\alpha^2}\right]}$

Solution:

Head Supplied H = Work Done + Kinetic energy head at exit.

Therefore, $H = \dfrac{V_2^2}{2g} + \dfrac{1}{g}(V_{w1}u_1 \pm V_{w2}u_2)$; For maximum efficiency $V_{w2} = 0$,

Also $V_{f2} = V_2$ and $V_{f2} = kV_{f1}$

From inlet velocity triangle,

$V_{w1} = u_1$ and $V_{f1} = u_1 tan\alpha$

Therefore, $H = \dfrac{V_{w1}u_1}{g} + \dfrac{V_2^2}{2g} = \dfrac{u_1^2}{g} + \dfrac{k^2 u_1^2 tan_\alpha^2}{2g}$

Therefore, peripheral speed $u_1 = \sqrt{\left[\dfrac{2gH}{2+k^2 tan_\alpha^2}\right]}$

Problem 18:

(a) Show that the hydraulic efficiency for a Francis turbine having velocity flow through runner constant, is given by the relation,

$\eta_h = [1/(1 + (\dfrac{1}{2}tan_\alpha^2)/(1 - \dfrac{tan\alpha}{tan\theta})$

Where α is guide blade angle, θ is runner vane angle at outlet.

(b) If the vanes are radial at inlet, then show that,

$\eta_h = \dfrac{2}{2 + tan_\alpha^2}$

Solution:

Velocity of flow is constant, therefore $V_{f2} = V_{f1}$

Discharge is radial at outlet, therefore $V_{w2} = 0$ and $V_{f2} = V_2$

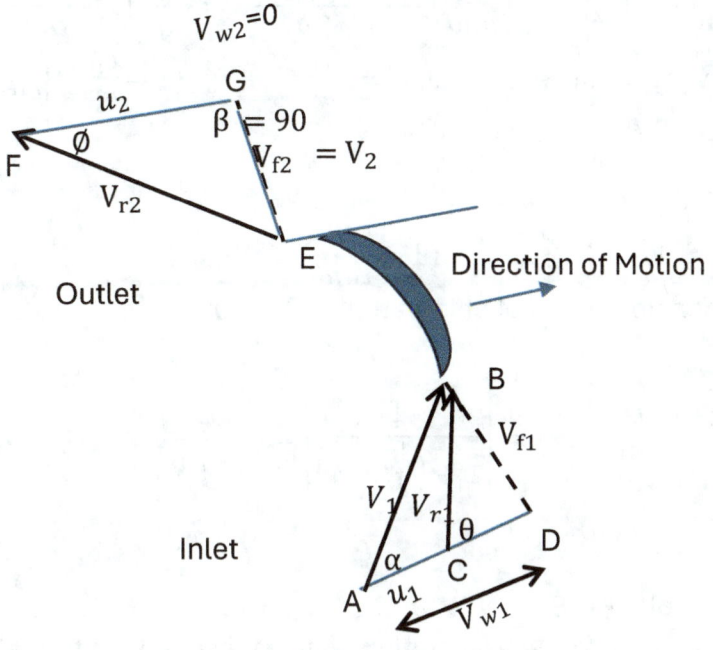

Figure: Problem 18.

From inlet velocity triangle, $tan\alpha = \frac{V_{f1}}{V_{w1}}$ or $V_{f1} = V_{w1} tan\alpha$

Also $tan\theta = \frac{V_{f1}}{V_{w1}-u_1}$ or $V_{w1} - u_1 = \frac{V_{f1}}{tan\theta}$

$$or, V_{w1} - u_1 = \frac{V_{w1}}{tan\theta} tan\alpha$$

$$Therefore \; u_1 = V_{w1}[1 - \frac{tan\alpha}{tan\theta}]$$

Head under which turbine is working:

$$H = \frac{V_{w1}u_1}{g} + \frac{V_2^2}{2g} = \frac{V_{w1}u_1}{g} + \frac{V_{f1}^2}{2g}$$

On substituting the values of V_{f1} and u_1, we get

$$H = \frac{V_{w1}}{g}V_{w1}\left(1 - \frac{tan\alpha}{tan\theta}\right) + \frac{(V_{w1}tan\alpha)^2}{2g} = \frac{V_{w1}^2}{g}[1 - \frac{tan\alpha}{tan\theta} + \frac{tan_\alpha^2}{2}]$$

Now hydraulic efficiency $\eta_h = \frac{V_{w1}u_1}{gH} = [V_{w1}^2(1 - \frac{tan\alpha}{tan\theta})]/[g.$

$\frac{V_{w1}^2}{g}(1 - \frac{tan\alpha}{tan\theta} + \frac{1}{2}tan_\alpha^2)$

Therefore

$$\eta_h = [1/(1 + (\frac{1}{2}tan_\alpha^2)/(1 - \frac{tan\alpha}{tan\theta})$$

When flow is radial at inlet that is $\theta = 90^o$

Then

$$\eta_h = \left[\frac{1}{1 + \left(\frac{1}{2}tan_\alpha^2\right)}\right] = \frac{2}{2 + tan_\alpha^2}$$

Problem 19:

In a hydro-electric station water is available at the rate of 175 $\frac{m^3}{s}$, under a head of 18 m. The turbines run at a speed of 150 rpm with overall efficiency of 82%. Find the number of turbines required if they have maximum specific speed of 460.

Solution :

$$N_s = \frac{N\sqrt{P}}{H^{\frac{5}{4}}} \text{ or } P = \left[\left(\frac{460x(18)^{\frac{5}{4}}}{150}\right)^2\right] = 12927.5\ kw$$

Power available for turbine $= \eta_o$

$\rho g Q H = 9.81x175x18x0.82 = 25339.23\ kw$

Therefore, number of turbines required=25339.23/12927.5 = 2 (approximately).

Problem 20:

A Francis turbine has been manufactured to develop 15000 hp at the head of 81 m and speed 375 rpm. The mean atmospheric pressure at site is 1.03 Kgf/cm^2 and vapour pressure at site is 0.03 Kgf/cm^2. Calculate the

92

maximum possible height of the runner above the tail water level to ensure cavitation free operation. The critical cavitation factor for Francis turbine is given by $\sigma_c = 317x10^{-18}xN_s^2$, where N_s is the specific speed of the turbine in MKS units.

Solution:

$P = 15000, H = 81\ m, N = 375\ rpm,$

$Atmospheric\ Pressure\ p_a = 1.03\ \dfrac{Kgf}{cm^2}$

$= 1.03x9.81x10^4(\dfrac{N}{m^2})$

$Atmospheric\ Pressure\ head\ H_{atm} = \dfrac{p_a}{\rho g}$

$= \dfrac{1.03x9.81x10^4}{1000x9.81} = 10.3\ m$

$Vapour\ Pressure\ haed\ H_v = \dfrac{0.03x9.81x10^4}{1000x9.81} = 0.3\ m$

$Specific\ speed\ of\ turbine N_s = \dfrac{N\sqrt{P}}{H^{\frac{5}{4}}} = 189\ rpm$

$Critical\ cavitation\ factor\ \sigma_c$

$= 317x10^{-18}xN_s^2;\ N_s\ is\ in\ MKS\ (rpm, hp, m)$

$\sigma_c = 317x10^{-8}.N_s^2 = 0.1132$

$\sigma_c = \dfrac{H_{atm} - H_v - H_s}{H}\ or\ 0.1132 = \dfrac{10.3 - 0.3 - H_s}{81}$

Where H_s is the suction pressure head at outlet or height of turbine runner above tail water race.

$Therefore\ H_s = 10 - 0.113x81 = 0.8308$

Hence the maximum permissible height is 0.83 m above the tail water level.

Chapter 3: Centrifugal Pumps:

The hydraulic machines, which converts the mechanical energy into hydraulic energy are called pumps. The hydraulic energy is in the form of pressure energy. If the mechanical energy is converted by means of centrifugal force acting on the fluid, the hydraulic machine is called Centrifugal Pump.

The centrifugal pump acts as a reverse of an inward radial flow reaction turbine. This means that the flow in centrifugal pumps is in the radial outward directions.

The centrifugal pump works on the principle of forced vortex flow, which means that when a certain mass of liquid is rotated by an external torque, the rise in pressure head of rotating liquid takes place.

The rise in pressure head at any point of rotating liquid is proportional to the square of tangential velocity of the liquid at that point.

That is rise in pressure head $= \dfrac{V^2}{2g}$ or $\dfrac{\omega^2 r^2}{2g}$.

3.1

Thus, at the outlet of the impeller, where radius is more, the rise in pressure head will be more and the liquid will be discharged at the outlet, with a high-pressure head. Due to this high-pressure head, the liquid can be lifted to a high level.

Main Parts of a Centrifugal Pump:

The following are the main parts of a Centrifugal pump.

1. Impeller.
2. Casing.
3. Suction pipe with a foot valve.
4. Delivery pipe.

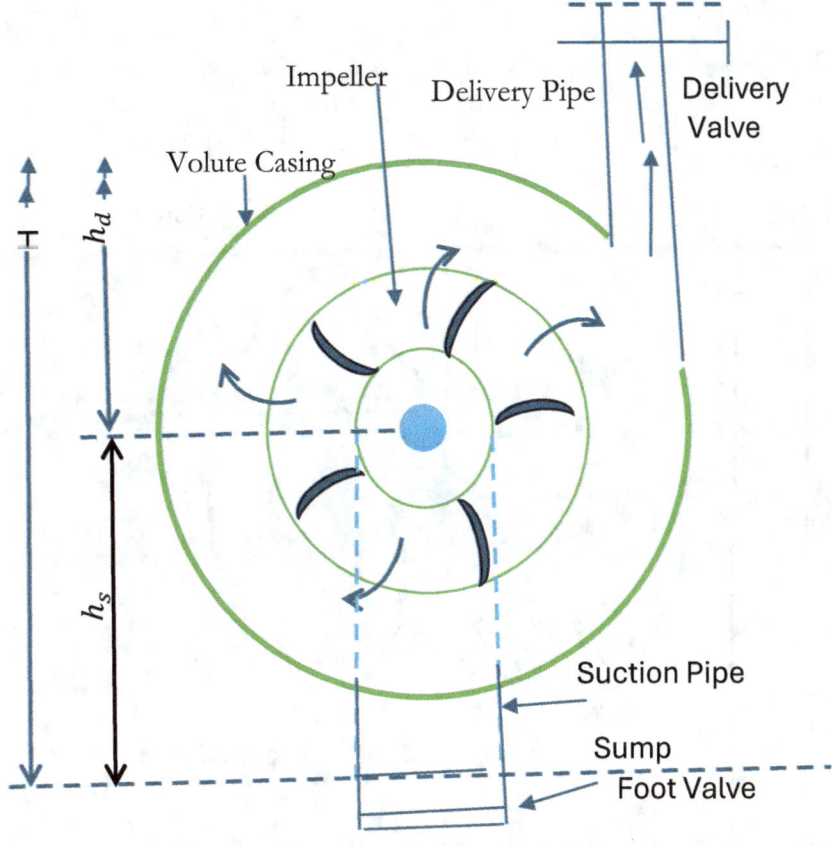

Figure 24: Main Parts of a Centrifugal Pump.

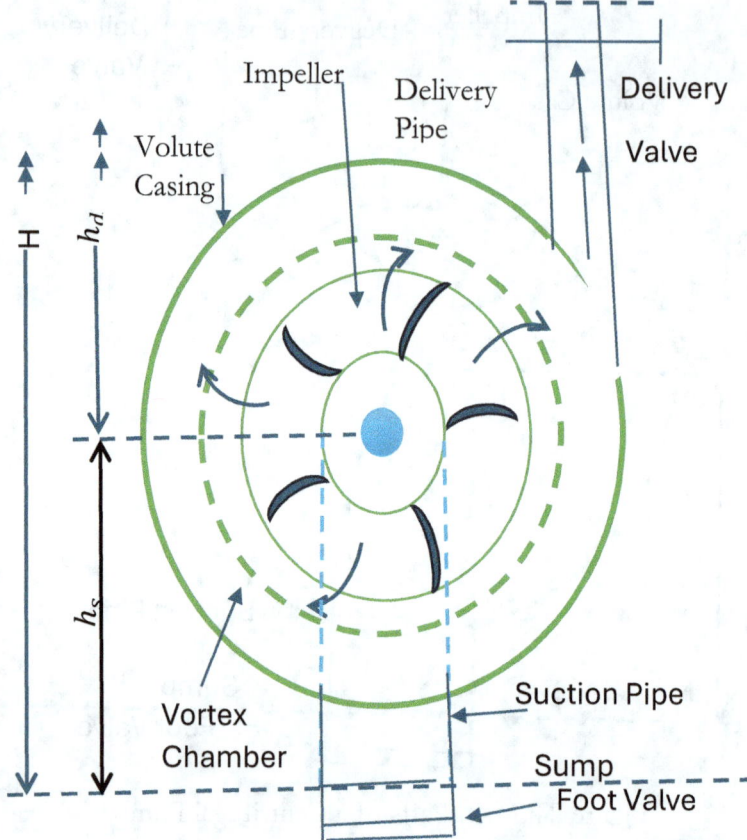

Figure 25: Centrifugal Pump with vortex chamber

Work done by the Centrifugal Pump (or by Impeller):

The expression for the work done by the impeller on the water is obtained by drawing velocity triangles at inlet and outlet of the impeller.

96

The water enters the impeller radially at inlet, for best efficiency of the pump, which means the absolute velocity of water makes an angle of 90° with the direction of motion of the impeller at the inlet.

Hence angle $\alpha = 90^o$ and $V_{w1} = 0$.

The figure shows the velocity triangles at the inlet and outlet tips of the vanes fixed to the impeller.

Let   N = Speed of impeller in rpm.

D_1 = Diameter of impeller at inlet.

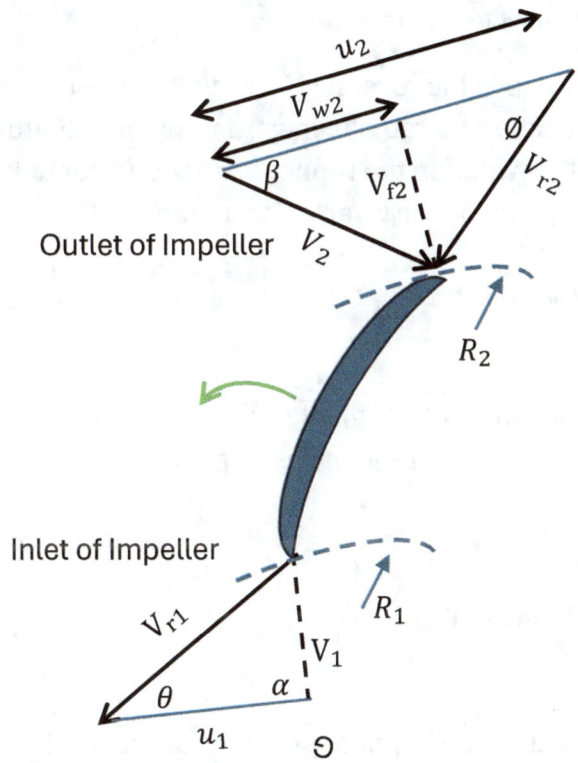

Figure 27: Velocity Triangles at inlet and outlet of Centrifugal Pump.

u_1 = Tangential velocity of impeller at inlet. $= \dfrac{\pi D_1 N}{60}$

D_2 = Diameter of impeller at outlet.

u_2 = Tangential velocity of impeller at inlet. $= \dfrac{\pi D_2 N}{60}$

97

V_1 = Absolute velocity of water at inlet.

V_{r1} = Relative velocity of water at inlet.

α = Angle made by absolute velocity (V_1) at inlet with the direction of motion of vane.

θ = Angle made by relative velocity (V_{r1}) at inlet with the direction of motion of vane.

V_2, V_{r2}, β and φ are the corresponding values at the outlet.

As the water enters the impeller radially, which means the absolute velocity of water at inlet is in the radial direction and hence $\alpha = 90^o$ and $V_{w1} = 0$.

A centrifugal pump is the reverse of a radially inward flow reaction turbine. But in case of a radially inward flow reaction turbine, the work done by the water on the runner per second per unit weight of water striking per second is given by equation (2.29),

$$= \frac{1}{g}[V_{w1}.u_1 - V_{w2}.u_2]$$

3.2

Therefore, work done by the impeller =
$$-[Work\ done\ in\ case\ of\ Turbine]$$

$$= -\frac{1}{g}[V_{w1}.u_1 - V_{w2}.u_2]$$

$$= \frac{1}{g}[V_{w2}.u_2] \; ; \text{Because } V_{w1} = 0$$

3.3

Therefore, work done by impeller on water per second,

$$= \frac{W}{g}[V_{w2}.u_2]$$

3.4

Where, $W = \rho g Q$, where Q is the volume of water striking impeller per second.

Also, Q = Area of flow x Velocity of flow

$$Q = \pi D_1 B_1 V_{f1} = \pi D_2 B_2 V_{f2}$$

3.5

Where B_1 and B_2 are width of impeller and V_{f1} and V_{f2} are velocity of flow at inlet and outlet.

Equation (2.3) gives the head imparted to the water by the impeller or energy given by impeller to water per unit weight per second.

Definitions of Heads and Efficiencies of Centrifugal Pump:

1. Suction Head (h_s):
   It is the vertical height of the centre line of the centrifugal pump above the water surface in the tank or pump from which water is to be lifted as shown in figure. This height is also called suction lift and is denoted as $'h_s'$.

2. Delivery Head (h_d):
   The vertical distance between the centre line of the pump and the water surface in the tank to which water is delivered is known as delivery head. This is denoted by $'h_d'$.

3. Static Head (H_s):
   The sum of suction head and delivery head is known as static head.

$$H_s = h_s + h_d$$

3.6

4. Manometric Head (H_m):
   The manometric head is defined as the head against which a centrifugal pump has to work.
   It is given by the following expressions:

   (a) $H_m = Head\ imparted\ by\ the\ impeller\ to\ the\ water$
   $- Loss\ of\ head\ in\ pump.$

$$= \frac{1}{g}[V_{w2}.u_2] - loss\ of\ head\ in\ impeller\ and\ casing.$$

$$= \frac{1}{g}[V_{w2}.u_2] \qquad if\ losses\ are\ zero\ or$$

negligible in the pump.

3.7

(b) $H_m = Total\ head\ at\ outlet\ of\ pump$
$$- Total\ head\ at\ inlet\ of\ pump.$$

$$= \left[\frac{p_o}{\rho g} + \frac{V_o^2}{2g} + Z_o\right] - \left[\frac{p_i}{\rho g} + \frac{V_i^2}{2g} + Z_i\right]$$

3.8

Where, $\frac{p_o}{\rho g}$ = Pressure head at outlet of pump = h_d.

$\frac{V_o^2}{2g}$ = Velocity head at outlet of pump.

Z_o = vertical height of the outlet of the pump from the datum line and

$\frac{p_i}{\rho g}, \frac{V_i^2}{2g}, Z_i$ = Corresponding values at inlet of the pump i.e.

$h_s, \frac{V_s^2}{2g}$ and Z_s respectively.

(c)$H_m = h_s + h_d + h_{fs} + h_{fd} + \frac{V_d^2}{2g}$

3.9

Where, h_{fs} = Friction head loss in suction pipe.

h_{fd} = Friction head loss in delivery pipe.

V_d = Velocity of water in delivery pipe.

Efficiencies:

In case of a centrifugal pump, the power is transmitted from the shaft of the electric motor to the shaft of pump and then to impeller. From the impeller, the power is given to the water.

The following are the important efficiencies of ta centrifugal pump.

(a) Manometric Efficiency (η_{man}):

The ratio of the manometric head to the head imparted by the impeller to the water is known as manometric efficiency.

$$\eta_{man} = \frac{Manometric\ head}{Head\ imparted\ by\ impeller\ to\ water}$$

$$= \frac{H_m}{\frac{V_{w2}.u_2}{g}} = \frac{gH_m}{V_{w2}.u_2}$$

3.10

Therefore, the ratio of the power given to water at the outlet of the pump to the power available at the impeller is known as manometric efficiency.

$$\eta_{man} = \frac{The\ power\ given\ to\ water\ at\ outlet\ of\ pump}{The\ power\ at\ impeller}$$

$$= \frac{\frac{WH_m}{1000}\ kw}{(Work\ done\ by\ impeller\ per\ second/1000)kw}$$

Therefore, $\eta_{man} = (\frac{WH_m}{1000})/(\frac{W}{g})(V_{w2}.u_2)/1000 = \frac{gH_m}{V_{w2}.u_2}$

3.11

(b) Mechanical Efficiency (η_m):

The ratio of the power available at the impeller to the power at the shaft of the centrifugal pump is known as mechanical efficiency.

$$\eta_m = \frac{Power\ at\ impeller}{Power\ at\ the\ shaft}$$

$$\eta_m = (\frac{W}{g})(V_{w2}.u_2)/(S.P.)$$

3.12

Where S.P. is shaft power

(c) Overall efficiency (η_o):

It is defined as ratio of power output of the pump to the power input to the pump.

$$\eta_o = \frac{The\ power\ output\ of\ the\ pump\ in\ KW}{Power\ input\ to\ the\ pump}$$

$$\eta_o = \frac{[\frac{(WH_m)}{1000}]}{Power\ supplied\ by\ the\ electric\ motor}$$

$$\eta_o = \frac{[\frac{(WH_m)}{1000}]}{Shaft\ Power\ of\ the\ pump}$$

$$\eta_o = \frac{[\frac{(WH_m)}{1000}]}{S.P.\ of\ the\ pump}$$

3.13

Or, $\eta_o = \eta_{man} x \eta_m$

3.14

Maximum Suction Lift (or Suction Height):

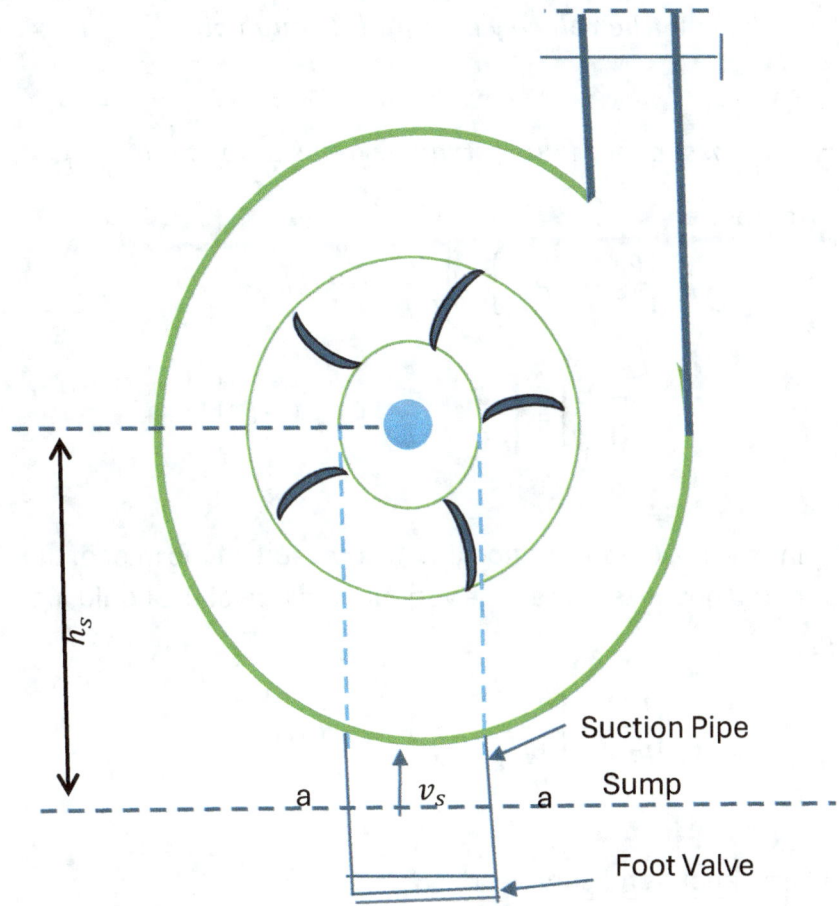

Figure 28: Maximum Suction Height of a Centrifugal Pump.

Figure 28, shows a layout of a centrifugal pump. Applying Bernoulli equation at the free surface of liquid in sump and section 1-1 in the suction pipe just at the inlet of the pump. Taking the free surface as datum line a-a.

$$\left[\frac{p_a}{\rho g} + \frac{V_a^2}{2g} + Z_a\right] = \left[\frac{p_1}{\rho g} + \frac{V_1^2}{2g} + Z_1 + h_L\right]$$

3.15

Where, $V_a = 0$, at the free of liquid.

$Z_a = 0$, the datum.

$V_1 = v_s$ the velocity of liquid through pipe.

$Z_1 = h_s$.

h_L = head loss in foot valve, strainer and suction pipe = h_{fs}.

Therefore,

$$\left[\frac{p_a}{\rho g} + 0 + 0\right] = \left[\frac{p_1}{\rho g} + \frac{v_s^2}{2g} + h_s + h_{fs}\right]$$

Or,

$$\left[\frac{p_1}{\rho g}\right] = \left[\frac{p_a}{\rho g} - \left[\frac{v_s^2}{2g} + h_s + h_{fs}\right]\right]$$

3.16

For finding the maximum suction lift, the pressure at the inlet of the pump should not be less than the vapour pressure of the liquid, i.e. $p_1 = p_v$.

$$\left[\frac{p_v}{\rho g}\right] = \left[\frac{p_a}{\rho g} - \left[\frac{v_s^2}{2g} + h_s + h_{fs}\right]\right]$$

3.17

Also, $\left[\frac{p_v}{\rho g}\right]$ = Vapour pressure head = H_v

& $\left[\frac{p_a}{\rho g}\right]$ = Atmospheric pressure head = H_a

Equation (2.17) can be written as,

$$H_a = H_v + \frac{v_s^2}{2g} + h_s + h_{fs}$$

Also,

$$h_s = H_a - H_v - \frac{v_s^2}{2g} - h_{fs}$$

3.18

If suction height is more, then vaporization of liquid at inlet will take place and it may give rise to cavitation.

Minimum Speed for Starting a Centrifugal Pump:

If the pressure rise in the impeller is more than or equal to manometric head (H_m), the centrifugal pump will start delivering water.

In case of a forced vortex, the centrifugal head or head due to pressure rise in the impeller,

$$= \frac{\omega^2 r_2^2}{2g} - \frac{\omega^2 r_1^2}{2g} = \frac{u_2^2}{2g} - \frac{u_1^2}{2g}$$

3.19

The flow of water will commence only if

$$\frac{u_2^2}{2g} - \frac{u_1^2}{2g} \geq H_m$$

For minimum speed

$$\frac{u_2^2}{2g} - \frac{u_1^2}{2g} = H_m$$

3.20

Also, from equation (3.11)

$$\eta_{man} = \frac{gH_m}{V_{w2}.u_2}$$

Or, $H_m = \eta_{man} \dfrac{(V_{w2}.u_2)}{g}$

Therefore,

$$\frac{u_2^2}{2g} - \frac{u_1^2}{2g} = \eta_{man} \frac{(V_{w2}.u_2)}{g}$$

105

Also, $\quad \frac{1}{2g}\left[\left(\frac{\pi D_2 N}{60}\right)^2 - \left(\frac{\pi D_1 N}{60}\right)^2\right] = \eta_{man}\frac{(V_{w2}.u_2)}{g} = \eta_{man}\frac{(V_{w2})\pi D_2 N}{g}\frac{}{60}$

$$\frac{\pi N}{120}\left[D_2^2 - D_1^2\right] = \eta_{man}V_{w2}D_2$$

Or, $\qquad\qquad N = \frac{[120.\eta_{man}.V_{w2}D_2]}{[\pi(D_2^2 - D_1^2)]}$

3.21

Equation (2.21) gives the minimum starting speed of a centrifugal pump.

Net Positive Suction Head (NPSH):

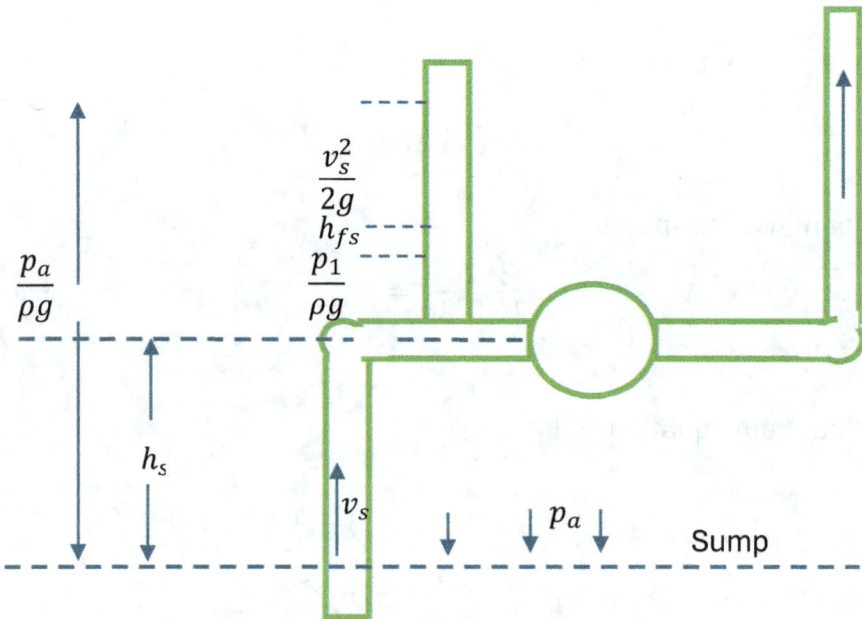

Figure 29: Pressure Balance at Pump Section.

The figure 29 shows pressure balance at pump section. The pump will work without cavitation, if p_1 is greater than p_v (vapour pressure of liquid for a given temperature) by an amount equal to that required by the liquid for the increase in velocity head when entering the impeller, if this amount be denoted by H_{sv}, we can write

$$\frac{p_1}{\rho g} = \frac{p_v}{\rho g} + H_{sv}$$

3.22

Also, from equation (2.16),

$$\left[\frac{p_1}{\rho g}\right] = \left[\frac{p_a}{\rho g} - \left[\frac{v_s^2}{2g} + h_s + h_{fs}\right]\right]$$

3.23

From equations (2.22) & (2.23), we get

$$H_{sv} = \frac{p_1}{\rho g} - \frac{p_v}{\rho g} = \left[\frac{p_a}{\rho g} - \left[\frac{v_s^2}{2g} + h_s + h_{fs}\right]\right] - \frac{p_v}{\rho g}$$

Also, $H_a = \frac{p_a}{\rho g}$ & $H_v = \frac{p_v}{\rho g}$

Therefore, $H_{sv} = H_a - H_s - H_v$

3.24

Where $H_s = Total\ suction\ head = \left[\frac{v_s^2}{2g} + h_s + h_{fs}\right]$

The value of H_{sv} is frequently called the net positive suction head (NPSH).

Thus, NPSH may be defined as "the net head (in meters of liquid) that is required to make liquid flow through the suction pipe from the sump to the impeller".

Priming of a Centrifugal Pump:

Priming of a centrifugal pump is defined as the operation in which the suction pipe, casing of the pump and a portion of the delivery pipe up to the delivery valve is completely filled up from outside source with the liquid to be raised by the pump before starting the pump. Thus, air from these parts of the pump is removed and these parts are filled with the liquid to be pumped.

Head generated by pump $= \dfrac{(V_{w2}.u_2)}{g}$ meters. This equation is independent of density of the liquid. This means that when pump is running in air, the head generated is in terms of meter of air, which is negligible and hence water may not be sucked from the pump. To avoid this difficulty, priming is necessary.

Multi-stage Centrifugal Pumps:

If a centrifugal pump consists of two or more impellers, the pump is called multistage centrifugal pump. The impellers may be mounted on the same shaft or on different shafts.

A multistage has the following two important functions.

1. To produce a high head or
2. To discharge a large quantity.

If a high head is to be developed, the impellers are mounted in series (or on the same shaft) while for discharge a large quantity of liquid the impellers (or pumps) are connected in parallel.

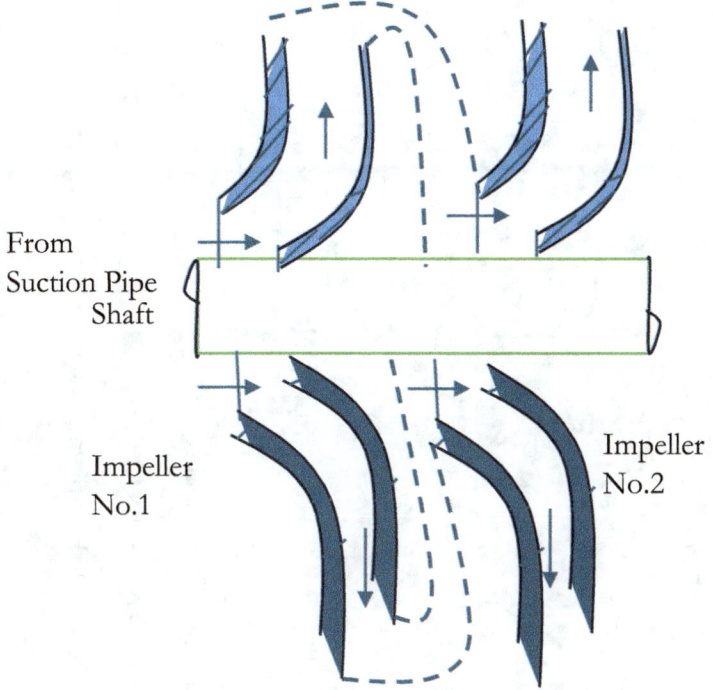

From
Suction Pipe
Shaft

Impeller
No.1

Impeller
No.2

Pipe connecting outlet of 1st. impeller to the inlet of 2nd impeller.
Let n = Number of identical impeller mounted on the same shaft & H_m
= Head developed by each impeller, then total head developed= $n\,H_m$.

Figure 30: Two stage pumps with impellers in series for high head.

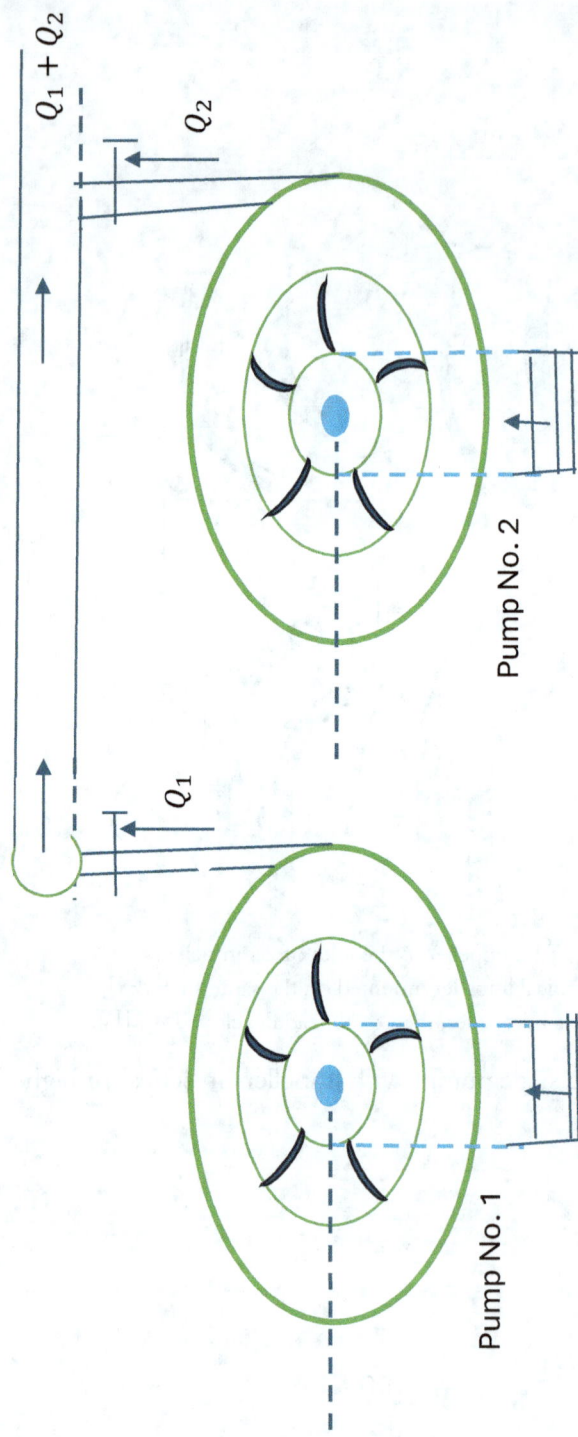

$Q_1 + Q_2$

Q_2

Q_1

Pump No. 1

Pump No. 2

Let n = Number of identical pumps arranged in parallel &

Q= Discharge from pump, then total discharge = n x Q.

Figure 31: Pumps in Parallel.

Specific Speed of a Centrifugal Pump(N_s):

The specific speed of a centrifugal pump is defined as the speed of a geometrically similar pump, which would deliver one cubic meter of liquid per second against a head of one meter. It is denoted by $'N_s'$.

Discharge Q = Area x Velocity of flow.

$$= \pi D B V_f \ or \ Q \propto D B V_f \ ; Also \ B \propto D \quad ; \text{Where D is the}$$
diameter of impeller & B width of impeller.

Therefore, $Q \propto D^2 V_f$

3.25

Tangential velocity $u = \dfrac{\pi D N}{60} \ or \ u \propto DN$

Also $u \propto V_f \propto \sqrt{H_m}$

Therefore, $\sqrt{H_m} \propto DN \ or \ D \propto \dfrac{\sqrt{H_m}}{N}$

From equation (2.25), $Q \propto D^2 V_f \propto \dfrac{H_m}{N^2}.V_f \propto \dfrac{H_m}{N^2}.\sqrt{H_m}$

Thus $Q \propto \dfrac{H_m^{\frac{3}{2}}}{N^2}$

Or $Q = k.\dfrac{H_m^{\frac{3}{2}}}{N^2}$

3.26

If $H_m = 1$, and $Q = \dfrac{1m^3}{s}$, then $N = N_s$

Therefore, $k = N_s^2$

So, from equation (2.26),

$$Q = N_s^2 \dfrac{H_m^{\frac{3}{2}}}{N^2}$$

Therefore, $N_s^2 = \dfrac{Q}{\frac{H_m^{\frac{3}{2}}}{N^2}} = \dfrac{N^2 Q}{H_m^{\frac{3}{4}}} or \ N_s = N\sqrt{Q}/H_m^{3/4}$

Model Testing of Centrifugal Pump:

Before manufacturing the large size pumps, their models, which are in complete similarity with the actual pumps, also called prototypes are made. Tests are conducted on the models and performance of the prototypes are predicted.

The complete similarity between the model and actual pump (prototypes) is predicted.

The complete similarity between the model and actual pump (prototype) exists if the following conditions are satisfied.

1. Specific speed of model = Specific speed of prototype.

$$(N_s)_m = (N_s)_p \text{ or } (N\sqrt{Q}/H_m^{3/4})_m = (N\sqrt{Q}/H_m^{3/4})_p$$

3.28

2. Tangential velocity.

$$u \propto \sqrt{H_m} \propto DN$$

Therefore $\frac{\sqrt{H_m}}{DN} = constant$

$$\left(\frac{\sqrt{H_m}}{DN}\right)_m = \left(\frac{\sqrt{H_m}}{DN}\right)_p$$

3.29

3. Discharge.

$$Q \propto D^2 V_f \text{ and } V_f \propto u \propto DN$$

So, $Q \propto D^3 N$ or $\frac{Q}{D^3 N} = constant$

$$\left(\frac{Q}{D^3N}\right)_m = \left(\frac{Q}{D^3N}\right)_p$$

3.30

4. Power of pump.

$$P = \frac{\rho g Q H_m}{75} \ or \ P \propto Q H_m \propto D^3 N.D^2 N^2$$

Therefore $\frac{P}{D^5N^2} = constant$

Or

$$\left(\frac{P}{D^5N^2}\right)_m = \left(\frac{P}{D^5N^2}\right)_p$$

3.31

Characteristics Curves of Centrifugal Pumps:

Characteristic curves of centrifugal pumps are defined as those curves which are plotted from the results of a number of tests on the centrifugal pump. These curves are necessary to predict the behaviour and performance of the pump when the pump is working under different flow rates, head and speed. The following are the important characteristic curves for pumps.

1. Main Characteristic Curves: Main characteristic curves of a centrifugal pump consists of variation of head (manometric head) vs speed, discharge is kept constant. For plotting curves of discharge vs speed, manometric head(H_m) is kept constant. And for plotting curves of power vs speed, the manometric head and discharge are kept constant.

    For plotting the graph of H_m vs speed (N), the discharge is kept constant.

    From the specific speed relation, equation (2.28), it is clear that $\frac{\sqrt{H_m}}{DN} = constant. \ or \ H_m \propto N^2$. It means that the head developed

113

by a pump is proportional to N^2. Hence curve of H_m vs speed (N) is a parabolic curve.

From relation, equation (2.31), $\frac{P}{D^5 N^3}$ is constant.

Hence $P \propto N^3$. This means that the curve P vs N is a cubic curve.

From relation of equation (2.30), $\frac{Q}{D^3 N} = contant.$

It means Q is proportional to N for a given pump. Hence the curve Q vs N is a straight line as shown in figure 32.

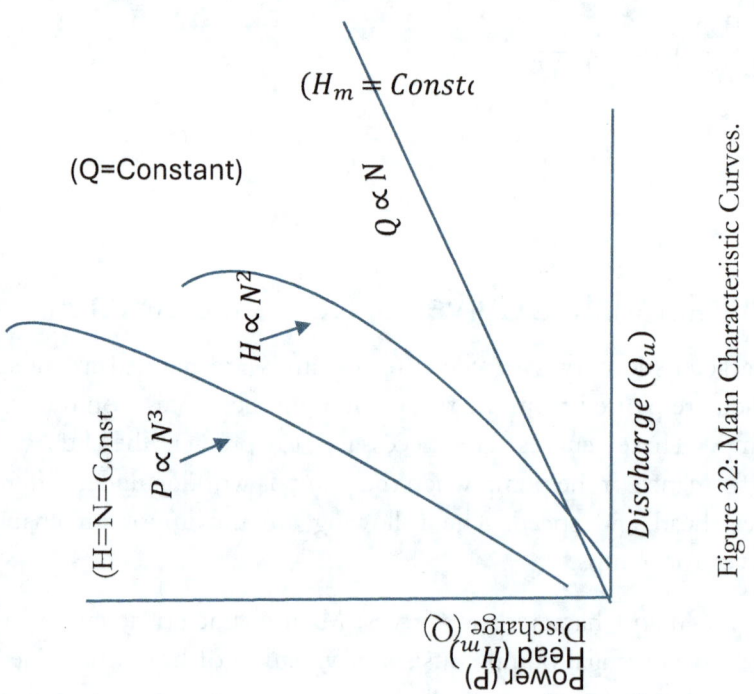

Figure 32: Main Characteristic Curves.

2. Operating Characteristic Curves: If the speed is kept constant, the variation of manometric head, power and efficiency with respect to discharge gives the operating characteristics of the pump.

The input power curve for pumps shall not pass through the origin. It will be slightly away from the origin on the y-axis, as even at zero discharge some power is needed to overcome mechanical losses.

The head curve will have maximum value of head when discharge is zero. The output power curve will start from origin as at Q=0, the output power (ρQgH) will be zero.

The efficiency curve will start from origin as at Q=0, η=0.

(Because, $\eta = \dfrac{\text{Output}}{\text{Input}}$).

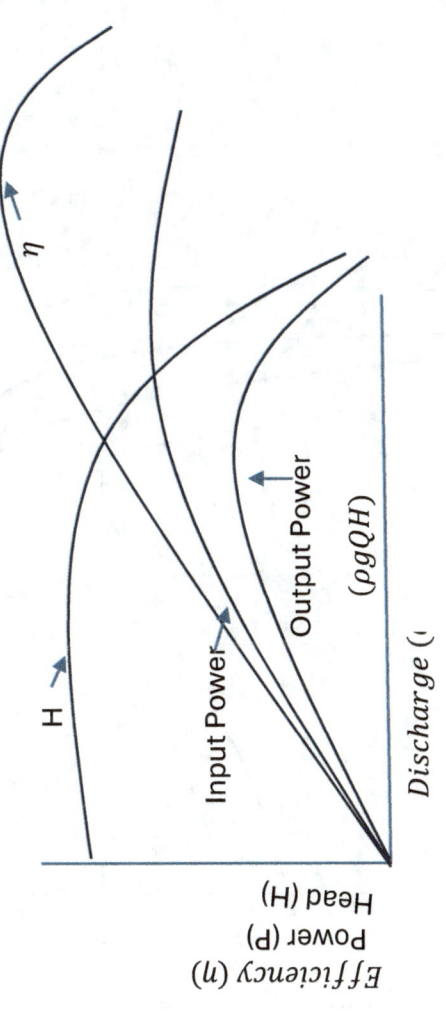

Figure 33: Operating Characteristic Curves.

3. Constant Efficiency Curves: For obtaining constant efficiency curves for a pump, the head vs discharge curves for different speeds are used.

For plotting the constant efficiency (Iso-efficiency) curves, horizontal lines representing constant efficiencies are drawn on the η-Q curves. The points, at which these lines cut the efficiency curves at various speeds, are transferred to the corresponding H-Q curves

The points having the same efficiency are then joined by smooth curves. These smooth curves represent the iso-efficiency curves.

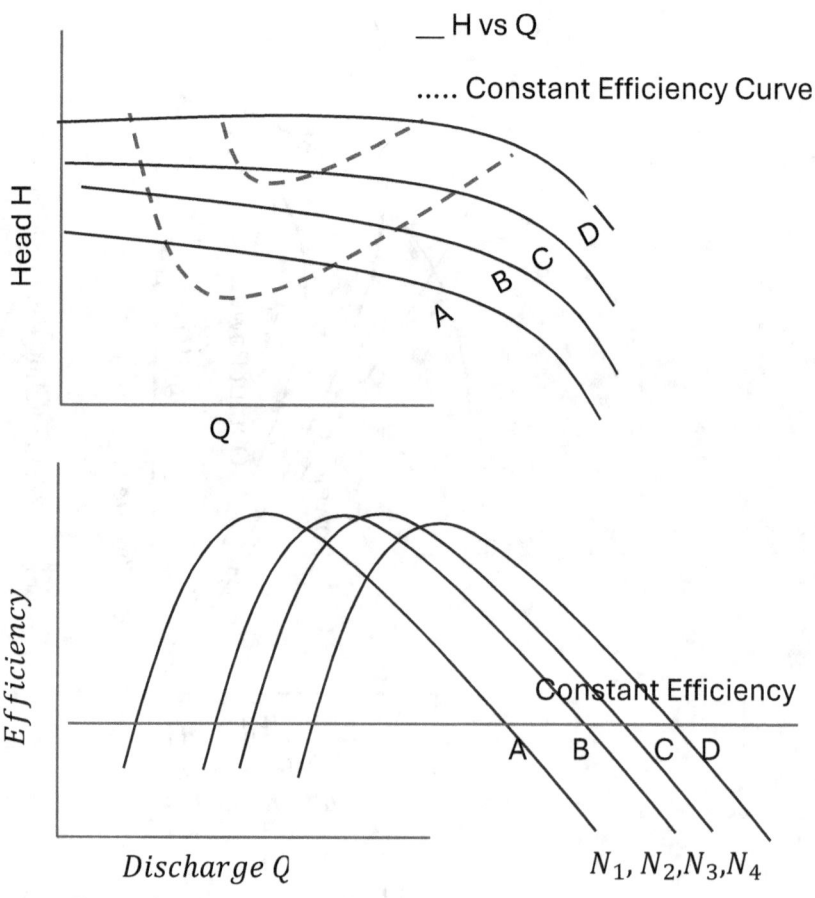

Figure 34: Constant Efficiency Curve.

Cavitation:

Cavitation is defined as the phenomenon of formation of vapour bubbles of a flowing liquid in a region where the pressure falls below its vapour pressure and the sudden collapsing of these vapour bubbles in a region of higher pressure.

When the vapour bubbles collapse, a very high pressure is created. The metallic surfaces, above which these bubbles collapse, are subjected to these high pressures which cause pitting action on the

117

surface. Thus, cavities are formed on the metallic surface and also considerable noise and vibrations are produced.

Effected regions in a reaction turbine are inlet of runner and draft tube and those in case of centrifugal pump are suction side and inlet of impeller.

Effects of cavitation:

(i)   Surface damaged and cavities are formed.

(ii)   Noise & Vibration.

(iii)   Efficiency drops.

Precautions against cavitation:

(i)   Pressure of the flowing liquid in any part of the hydraulic system should not be allowed to fall below its vapour pressure. For water, absolute pressure should not be below 2.5 m of water.

(ii)   Special material or coating such as Al-bronze and stainless steel, which are cavitation resistant materials, should be used.

Cavitation in Turbines:

Professor Dietrich Thoma (of Munich Germany) suggested a cavitation factor to determine the zone where turbine can be operated without being affected from cavitation.

Thoma's cavitation factor, a dimensionless number.

$$\sigma = \frac{H_b - H_s}{H} = \frac{[(H_{atm} - H_v) - H_s]}{H}$$

3.32

Where,

H_b = Barometric pressure head in meters of water.

118

H_{atm} = Atmospheric pressure head in meters of water.

H_v = Vapour pressure head in meters of water.

H_s
= Suction pressure head at the outlat in meters of water or

height of Turbine runner above the water surface.

H = Net head on the turbine in meters of water.

Thoma's cavitation factor for centrifugal pump:

$$\sigma = \frac{H_b - H_s - h_{ls}}{H} = \frac{[(H_{atm} - H_v) - H_s - h_{ls}]}{H}$$

3.33

Where,

H_b = Barometric pressure head in meters of water.

H_{atm}
= Atmospheric pressure head in meters of water or absolute

Pressure head at liquid surface in pump.

H_v = Vapour pressure head in meters of water.

H_s
= Suction pressure head at the outlat in meters of water or

height of Turbine runner above the water surface.

H = Head developed by pump in meters of water.

h_{ls}
= Head lost due to friction in suction pipe in meters of water.

The values of cavitation factor $'\sigma'$ obtained from equations (2.32 & 2.33) are compared with critical cavitation factor $'\sigma_c'$ for the turbine or pump. If $> \sigma_c$, then cavitation will not occur.

119

Following empirical relationships for obtaining the value of $'\sigma_c'$ for different turbines:

For Francis Turbines, $\sigma_c = 0.625.(\frac{N_s}{380.78})^2 = 431.10^{-8}N_s^2$

3.34

For Propeller Turbines, $\sigma_c = 0.28 + [\frac{1}{7.5}.(\frac{N_s}{380.78})^3$

3.35

In the above expressions $'N_s'$ is in [rpm, kw, m]

If $'N_s'$ is in [rpm, hp, m], then equations (2.34 & 2.35) will become as follows:

For Francis Turbine, $\sigma_c = 0.625.(\frac{N_s}{444})^2 = 317.10^{-8}N_s^2$

3.36

For Propeller Turbine, $\sigma_c = 0.28 + [\frac{1}{7.5}.(\frac{N_s}{444})^3$

3.37

Cavitation in Centrifugal Pumps:

Thoma's cavitation factor for pump is

$$\sigma = \frac{[(H_{atm} - H_v) - H_s - h_{ls}]}{H} = \frac{[(H_a - H_v) - h_s - h_{fs}]}{H_m}$$

$$Where, H_s = h_s, h_{ls} = h_{fs}, H = H_m$$

$$Also, [(H_a - H_v) - h_s - h_{fs}] = NPSH$$

$$Therefore, \sigma = \frac{NPSH}{H_m}$$

If $\sigma < \sigma_c$ then Cavitation will not occur.

$$\sigma_c = 0.103(\frac{N_s}{1000})^{4/3} = 1.03x10^{-3}N_s^{4/3}$$

Water Hammer:

Water hammer (or more generally called Fluid hammer) ia a pressure surge or wave caused when a fluid (usually a liquid but sometimes also a gas) in motion is forced to stop or change direction suddenly (momentum change).

Water hammer commonly occurs when a valve is closed suddenly at an end of pipeline system and pressure wave propagates in the pipe. This pressure wave can cause major problems, from noise and vibration to collapse.

Worked Examples

Problem 21:

A centrifugal pump rotating at 1000 rpm delivers 160 litres/s of water against a head of 30 m. The pump is installed at a place where atmospheric pressure is $1x10^5 P_a(abs)$ and vapour pressure of water is $3\ kp_a(abs)$ .The head loss in suction pipe is equivalent to 0.2 m of water. Calculate minimum NPSH, and maximum allowable height of the pump from free surface of water in sump.

Solution:

$$p_a = 1x10^5 P_a = \frac{1x10^5 N}{m^2}; p_v = 3kP_a$$

$$= 3x10^3 \left(\frac{N}{m^2}\right); h_{fs} = 0.2m$$

$$\sigma = \frac{NPSH}{H_m}$$

When $\sigma = \sigma_c$, then NPSH will be minimum.

$$\sigma_c = \frac{(NPSH)_{min}}{H_m}, therefore \ (NPSH)_{min} = \sigma_c x H_m$$

$$\sigma_c = 1.03x10^{-3} x N_s{}^{4/3}$$

$$N_s = \frac{N\sqrt{Q}}{H_m^{\frac{3}{4}}} = \frac{1000x\sqrt{0.16}}{(30)^{\frac{3}{4}}}$$

$$\sigma_c = 1.03x10^{-3} x \left(\frac{1000x\sqrt{0.16}}{(30)^{\frac{3}{4}}}\right)^{4/3} = 0.1012$$

Therefore, $(NPSH)_{min} = \sigma_c x H_m = 0.1012x30$
$$= 3.036 \ m.$$

Value of suction head (h_s) will be maximum, if NPSH is minimum,

$$(NPSH)_{min} = H_a - H_v - (h_s)_{max} - h_{fs}$$

$$H_a = \frac{p_a}{\rho g} = \frac{1x10^5}{1000x9.81} = 10.193 \ m \ of \ water \ column.$$

$$H_v = \frac{p_v}{\rho g} = 0.305 \ m \ of \ water \ column, h_{fs}$$

$$= 0.2 \ m \ and \ (NPSH)_{min} = 3.036 \ m.$$

Therefore $(h_s)_{max} = 10.193 - 0.3036 - 0.2 - 3.036$
$$= 6.652 \ m.$$

Problem 22:

A three stage Centrifugal pump has an impeller 400 mm in diameter and 20 mm wide at outlet. The vanes are curved backward at the outlet at $45°$ and reduce the circumferential area by 20%. The manometric efficiency is 90% and the overall efficiency is 80%. The pump is running at 1000 rpm and delivering 0.05 m³/s. Determine (i) head generated by the pump and (ii) shaft power required to run the pump.

Solution:

Number of stages, $n = 3, D_2 = 0.4\ m, B_2 = .02\ m, \varnothing = 45°$

,

$Reduction\ in\ area = 10\%,\ N = 1000\ rpm,\ Q$

$$= 0.05\ \frac{m^3}{s},\ \eta_{man} = 90\%,\ \eta_o = 80\%.$$

$Area\ of\ flow = 0.9 x \pi D_2 B_2 = 0.02262\ m^2$

$Velocity\ of\ flow\ at\ outlet\ V_{f2} = \dfrac{Q}{Area\ of\ flow}$

$$= \frac{.05}{.02262} = 2.21 \frac{m}{s}.$$

$$u_2 = \frac{\pi D_2 N}{60} = 20.94 \frac{m}{s}$$

$$tan\varnothing = \frac{V_{f2}}{u_2 - V_{w2}}\ or\ V_{w2} = u_2 - \frac{V_{f2}}{tan\varnothing}$$

$$= 20.94 - \frac{2.21}{tan45°} = 18.73 \frac{m}{s}.$$

$$\eta_{man} = \frac{gH_{man}}{V_{w2}u_2}\ or\ H_{man} = \frac{0.9 x 18.73 x 20.94}{9.81} = 35.98\ m$$

(i) $H_{total} = nxH_{man} = 107.94\ m.$

(ii) $Shaft\ Power\ Required.$

$Power\ output\ of\ pump = \dfrac{\rho g Q x H_{total}}{1000} = 52.94\ kw$

$Overall\ efficiency \eta_o = \dfrac{Power\ output}{Power\ input} = \dfrac{52.94}{P}$

$$= 0.8\ or\ P = 66.17\ kw$$

Problem 23:

It is required to pump water out of deep well under a total head of 90 m. A number of identical pumps of design speed 1000 rpm and specific speed 30 with a rated capacity of 0.15 m³/s are available. How many pumps are required and how should they be connected whether in series or parallel?

Solution:

$$N_s = \frac{N\sqrt{Q}}{H_{man}^{\frac{3}{4}}} \text{ or } 30 = \frac{1000\sqrt{0.15}}{H_{man}^{\frac{3}{4}}} \text{ or } H_{man}$$

$$= [\frac{1000\sqrt{0.15}}{30}]^{\frac{4}{3}} = 30.28 \ m$$

Number of stages $= \frac{H_{total}}{H_{man}} = \frac{90}{30.28} \cong 3$

As the total head is 90 m, and the head developed by each pump is about 30 m, so they will be connected in series.

Problem 24:

In order to predict the performance of a large centrifugal pump, a scale model of one-sixth size is made with the following specifications:

Power P=25 kw, Head $H_{man} = 7 \ m$, Speed N=1000 rpm. If the prototype pump has to work against a head of 22 m, calculate its working speed, power required to drive it and the ratio of the flow rates handled by two pumps.

Solution:

Scale ratio is one-sixth.

Model: Power (P_m) = 25 kw; Head (H_m) = 7 m; Speed (N_m)= 1000 rpm.

Prototype: Power (P_p) =? ;Head (H_m) = 22 m; Speed (N_m)= ?.

Speed of prototype (N_p):

$$\left(\frac{\sqrt{H_m}}{DN}\right)_m = \left(\frac{\sqrt{H_{man}}}{DN}\right)_p$$

$$Therefore,\ N_p = \frac{\left(\sqrt{H_{man}}\right)_p}{\left(\sqrt{H_m}\right)_m} x N_m x \frac{D_m}{D_p}$$

$$= \frac{\sqrt{22}}{\sqrt{7}} x 1000 x \left(\frac{1}{6}\right) = 295.47\ rpm$$

Power required to drive prototype P_p :

$$\left[\frac{P}{D^5 N^3}\right]_m = \left[\frac{P}{D^5 N^3}\right]_p$$

$$Therefore,\ P_p = P_m x \left(\frac{D_p}{D_m}\right)^5 x \left(\frac{N_p}{N_m}\right)^3 = 25 x \left(\frac{6}{1}\right)^5 x \left(\frac{295.47}{1000}\right)^3$$

$$= 5014.6\ kw$$

Ratio of flow rates $\frac{Q_p}{Q_m}$:

$$\left(\frac{Q}{D^3 N}\right)_m = \left(\frac{Q}{D^3 N}\right)_p,\ Therefore\ \frac{Q_p}{Q_m} = \left(\frac{6}{1}\right)^3 x \left[\frac{295.47}{1000}\right]$$

$$= 63.82$$

Problem 25:

A centrifugal pump is discharging 0.025 m³/s of water against a total head of 18 m. The diameter of the impeller is o.4 m and it is rotating at 1400 rpm. Calculate the head, discharge and ratio of powers of a geometrically similar pump of diameter 0.25 m when it is running at 2800 rpm.

Solution:

Centrifugal pump: Discharge Q_1=0.025 m³/s; Head H_{man}= 18 m; Diameter D_1=0.4 m; Speed N_1=1400 rpm.
Geometrically similar pump: Q_2=?; H_{man2}=?; D_2=0.25 m; N_2=2800 rpm.

Head H_{man2}:

$$[\frac{\sqrt{H_{man}}}{DN}]_1 = [\frac{\sqrt{H_{man2}}}{DN}]_2 \text{;} Therefore \ H_{man2}$$

$$= \left[\frac{\sqrt{18}x0.25x2800}{0.4x1400}\right]^2 = 28.125 \ m$$

Discharge Q_2:

$$\left[\frac{Q}{D^3N}\right]_1 = \left[\frac{Q}{D^3N}\right]_2 \text{;} Therefore \ Q_2 = \frac{0.025x0.25^3x2800}{0.4^3x1400}$$

$$= 0.0122 \ m^3/s$$

Ratio of power P_1/P_2 :

$$[\frac{P}{D^5N^3}]_1 = [\frac{P}{D^5N^3}]_2 \text{;} Therefore \frac{P_1}{P_2} = \frac{D_1^5N_1^3}{D_2^5N_2^3}$$

$$= \frac{(0.4^5x1400^3)}{(0.25^5x2800^3)} = 1.31$$

Problem 26:

Show that the rise of pressure in the impeller of a centrifugal pump, when friction and other losses in the impeller are neglected, is given by

$$\frac{1}{2g}[V_{f1}^2 + u_2^2 - V_{f2}^2 cosec^2 \emptyset]$$

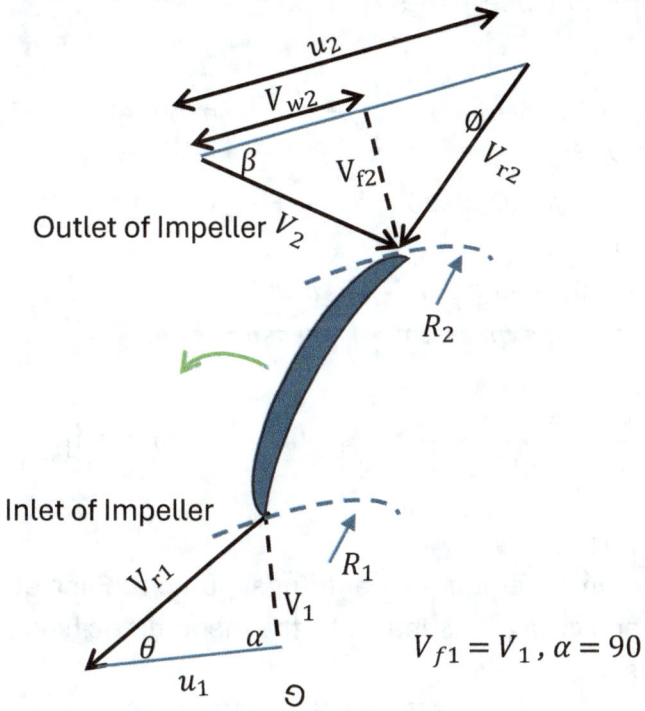

Outlet of Impeller V_2

Inlet of Impeller

$V_{f1} = V_1, \alpha = 90$

Figure: Problem 26.

Solution:
Applying Bernoulli's equation at the inlet and outlet of the impeller, neglecting losses from inlet to outlet, we get;

$$\left[\frac{p_1}{w} + \frac{V_1^2}{2g} + Z_1\right] = \left[\frac{p_2}{w} + \frac{V_2^2}{2g} + Z_2\right] - \text{work done by impeller}$$

on water per unit weight of water.

$$Or, \left[\frac{p_1}{w} + \frac{V_1^2}{2g} + Z_1\right] = \left[\frac{p_2}{w} + \frac{V_2^2}{2g} + Z_2\right] - \frac{V_{w2}u_2}{g}$$

If inlet and outlet of the pump are at the same level then $Z_1 = Z_2$.

So, the pressure rise in the impeller $= \dfrac{p_2}{w} - \dfrac{p_1}{w} = \dfrac{V_1^2}{2g} - \dfrac{V_2^2}{2g}$

$+ \dfrac{V_{w2}u_2}{g}$

From inlet velocity triangle, $V_1 = V_{f1}$; from outlet velocity triangle,

$V_{w2} = u_2 - V_{f2}cot\emptyset; also\ V_2^2 = V_{f2}^2 + V_{w2}^2 = V_{f2}^2 cosec^2 \emptyset + u_2^2 - 2u_2 V_{f2} Cot\emptyset$

Substituting the values of V_1, V_{w2} and V_2^2

in the expression of pressure rise,

we get

$$\dfrac{V_1^2}{2g} - \dfrac{V_2^2}{2g} + \dfrac{V_{w2}u_2}{g} = \dfrac{1}{2g}[V_{f1}^2 + u_2^2 - V_{f2}^2 cosec^2 \emptyset]$$

Problem no 27:

Show that, in general, for a centrifugal pump running at speed N and giving a discharge Q, the manometric head is expressible in the form

$$H_{mano} = AN^2 + BNQ + CQ^2$$

Where, A, B and C are constants.

Solution:

Manometric head is equal to the head imparted by the impeller minus the losses in the impeller and casing.

$$H_{mano} = \dfrac{V_{w2}u_2}{g} - K\dfrac{V_2^2}{2g}$$

(a)

$$u_2 = \dfrac{\pi D_2 N}{60} = K_1 N$$

$$V_{f2} = \dfrac{Q}{\pi D_2 B_2} = K_2 Q$$

$$V_{w2} = u_2 - V_{f2} Cot\emptyset = K_1 N - K_2 Q cot\emptyset = K_1 N - K_3 Q$$

From the outlet velocity triangle,

$$V_2^2 = V_{f2}^2 + V_{w2}^2 = (K_2 Q)^2 + (K_1 N - K_3 Q)^2$$

On substituting above values of V_{w2}, u_2 & V_2 in equation (a), we get

$$H_{mano} = AN^2 + BNQ + CQ^2$$

Chapter 4: Comparative Turbine Performance Evaluation

Performance Evaluation (Practical Approach)[21]

Normally characteristic curves are drawn for knowing the behavior and performance of the turbine under different working conditions. Various performance parameters are Speed (N), Head (H), Discharge (Q), Power (P), and overall efficiency. Out of these three are independent parameters (N, H & Q), one parameter is kept constant and variation of other parameters with respect to other two independent variables (say N and Q) are plotted, and various curves are obtained. In the present work, laboratory experimental data has been used to analyze waterpower utilized against the shaft power demand and variation of overall efficiency, specific speed, unit speed, unit power and unit discharge has been analyzed. All of the three Pelton, Francis and Kaplan small turbines have been used.

Experimental Set Up and Data

Water head is created by running a Centrifugal pump. Flow rate was measured by differential pressure across an orifice. Pressure head is measured by using a pressure gauge at inlet. Speed was measured by a Tachometer. Brake power was computed by using Rope Brake Dynamometer using pulley, rope, weight and spring balance. Manual governing was done.

Schematic of a Hydro Turbine in a Block Diagram

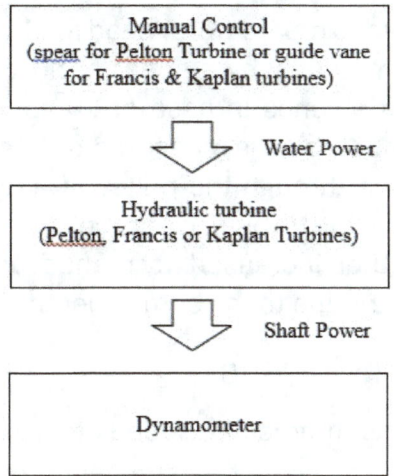

Figure 4.1: Block Diagram of Hydraulic Turbine Laboratory Facility.

Turbine Design Details

Turbine design details are given in the following table.

Table 4.2: Design Details of Small Turbines

Turbine	Pelton	Francis	Kaplan
Supply head m	46	15	5
Discharge lpm	80	1500	1500
Speed rpm	1500	1500	1500
Power Output Kw	3.75	3.75	3.75
Pitch circle diameter mm & Jet diameter mm	260 & 22	-	-
Runner inner/ outer diameter mm	-	127/ 160	110/ 200
Number of buckets/ vanes	18	8	4
Number of guide vanes	-	10	16

Methodology

A centrifugal pump is used to create water head for the turbine. The speed of turbine is set manually by a spear at nozzle in case of Pelton turbine or by guide vanes in case of Francis and Kaplan turbines. The speed of the turbine is measured by a tachometer. A resisting torque is applied through the pulley of a dynamometer which causes the speed of the turbine to reduce due to the applied load. The waterpower is then increased (by manual adjustment as above) to bring back the turbine to the same speed.

Discussions and Conclusions

During the laboratory experiments, the speed of turbines ranged from 1080 RPM to 1930 RPM, water head ranged from 5.5 m to 52 m and load varied from 0.53 KW to 1.9 KW for all the three turbines combined. Shaft power ratio (SPR), a non-dimensional quality is the ratio of shaft power demand and rated capacity 3.75 KW, which is same for all these Pelton, Francis and Kaplan small turbines and likewise Waterpower Ratio (WPR) is the ratio of water power utilized, to meet the shaft power demand, and rated capacity of turbines.

Figure 2 shows the variation of WPR with SPR for all the turbines. WPR increases with an increase in SPR and it exceeds 100% at SPR above 20%. It is due to overall efficiency being below 20% at lower demands in case of Pelton turbine, in case of Francis turbine, initially, it is constant up to 27% of shaft load, then it increases up to 36% of load and then it decreases. In Kaplan it has an increasing trend within the range of data analyzed.

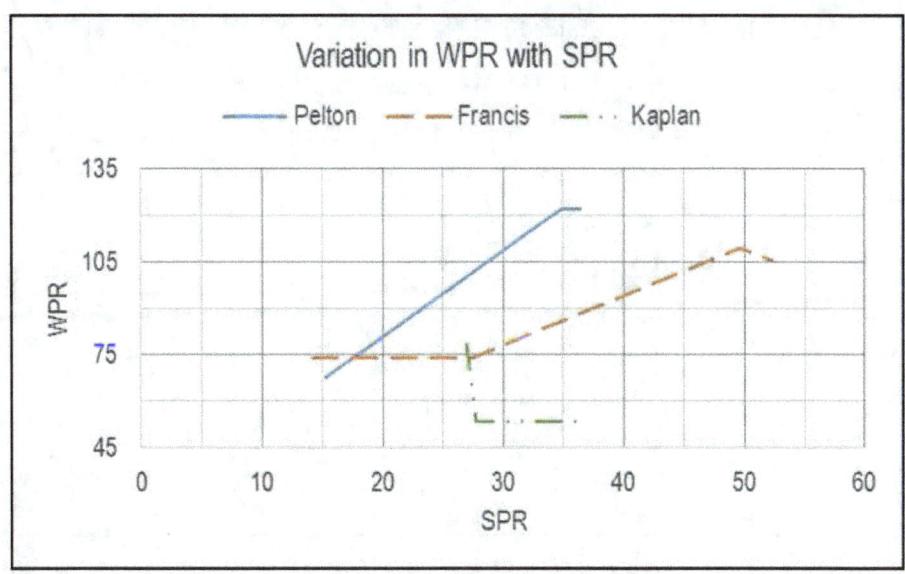

Figure 4.2: Variation of WPR with SPR.

Figure 4.3 shows the variation of SPR with specific speed. In case of Pelton turbine and Francis turbine, specific speed increases continuously with shaft power demand whereas for Kaplan turbine it has initial steep increase then slow creasing trend.

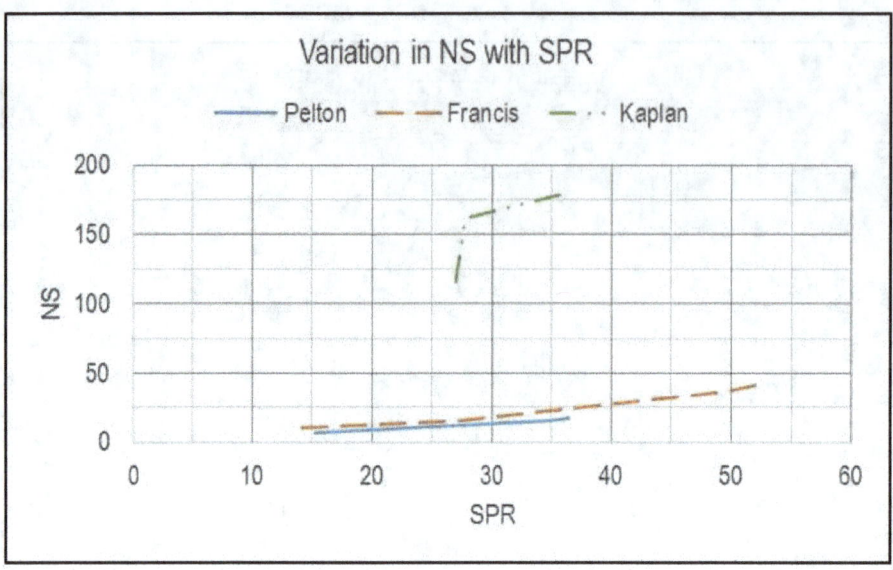

Figure 4.3: Variation of Specific Speed with SPR.

Figures 4.4 shows the variations of unit quantities with shaft power load (SPR). All the unit quantities have increasing trend with increasing shaft power demand (SPR) in all the turbines.

Unit speed has increasing trend in case of Pelton and Francis turbines as compared to that in Kaplan turbine. Unit power has an increasing trend in all the three turbines. Unit discharge has increasing trend in the case of Pelton turbine whereas comparatively it is steady in case of Francis and Kaplan turbines.

Overall unit quantities are not much affected with power demand changes in the case of Kaplan turbine.

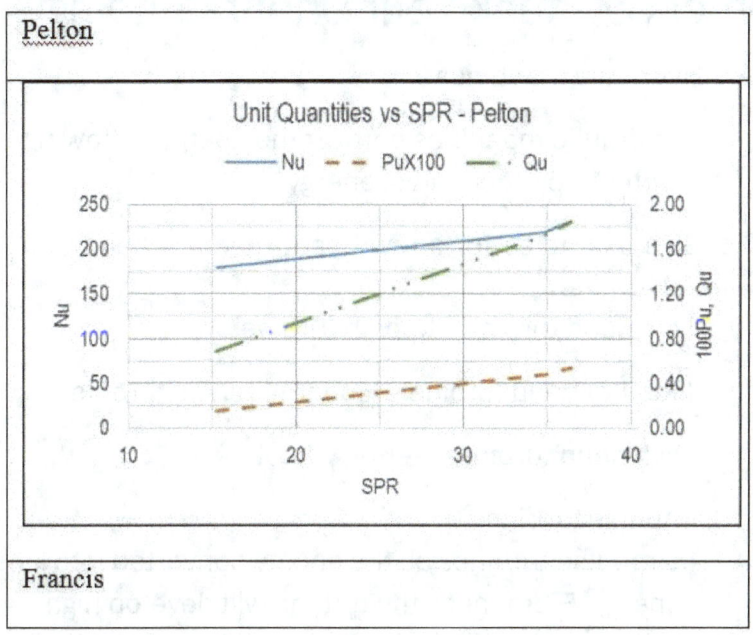

Figure 4.4: Variation of unit quantities with Shaft Power Load (SPR).

Maximum efficiency recorded during this analysis was 33.3 % for Pelton turbine, 49.5 % for Francis turbine and 68 % for Kaplan turbine.

This analysis is based on assorted data recorded during laboratory experiments for computation of overall efficiency of Pelton, Francis and Kaplan turbines. Further studies can be done to generate more data to arrive on trends and pattern of various turbine parameters.

Chapter 5: Check Your Understanding

1.Hydraulic machines:

Hydraulic machines convert the energy of flowing water into mechanical energy

2.Give examples for a low head, medium head and high head turbine.

Low head turbine – Kaplan turbine

Medium head turbine – Modern Francis turbine

High head turbine – Pelton wheel

3. Impulse turbine:

In impulse turbines all the energy converted into kinetic energy. From these the turbine will develop high kinetic energy power. This turbine is called impulse turbine.   Example: Pelton turbine

4.Reaction turbine:

In a reaction turbine, the runner utilizes both potential and kinetic energies. Here portion of potential energy is converted into kinetic energy before entering into the turbine.   Example: Francis and Kaplan turbine.

5.   Axial flow turbine:

In axial flow turbine water flows parallel to the axis of the turbine shaft.   Example: Kaplan turbine

6.Mixed flow turbine:

In mixed flow water enters the blades radially and comes out axially, parallel   to the turbine shaft. Example: Modern Francis turbine.

7.Spear and nozzle:

The nozzle is used to convert whole hydraulic energy into kinetic energy. Thus the nozzle delivers high speed jet. To regulate the water flow through the nozzle and to obtain a good jet of water spear or nozzle is arranged.

8. Heads:

Gross Head: The gross head is the difference between the water level at the reservoir and the level at the tailstock.

Effective Head: The head available at the inlet of the turbine.

9. Hydraulic efficiency:

It is defined as the ratio of power developed by the runner to the power supplied by the water jet.

10. Mechanical efficiency:

It is defined as the ratio of power available at the turbine shaft to the power developed by the turbine runner.

11. Volumetric efficiency:

It is defined as the volume of water actually striking the buckets to the total water supplied by the jet.

12. Overall efficiency:

It is defined as the ratio of power available at the turbine shaft to the power available from the water jet.

13. Pump:

A pump is a device which converts mechanical energy into hydraulic energy.

14. Main components of Centrifugal pump.

i) Impeller   ii) Casing   iii) Suction pipe, strainer & Foot valve   iv) Delivery pipe & Delivery valve

15. What is meant by Priming?

The delivery valve is closed and the suction pipe, casing and portion of the delivery pipe upto delivery valve are completely filled with the liquid so that no air pocket is left. This is called as priming.

16. Manometric head:

It is the head against which a centrifugal pump work.

17. Mechanical efficiency:

It is defined as the ratio of the power actually delivered by the impeller to the   power supplied to the shaft.

18. Overall efficiency:

It is the ratio of power output of the pump to the power input to the pump.

19. Speed ratio: It is the ratio of peripheral speed at outlet to the theoretical velocity of jet corresponding to manometric head.

20. Flow ratio: It is the ratio of the velocity of flow at exit to the theoretical velocity of jet corresponding to manometric head.

Multiple Choice Questions

Tick the correct answer

1. An ideal fluid is defined as the fluid which is
   (a) is compressible (b) is incompressible   (c) is incompressible and non-viscous   (d)has negligible surface tension.
2. Bernoulli's theorem deals with the law of conservation of
   (a) Mass and stiffness   (b) momentum (c) energy   (d) None of these

3. The specific speed of a turbine is defined as the speed of turbine which produces
   (a) unit power at unit head   (b) produces unit hp at unit discharge (c) delivers unit discharge at unit head   (d) delivers unit discharge at unit power.

4. The force exerted by a jet of water on a stationary vertical plate in the direction of jet is given by
   (a)$d.A.V^2 Sin\theta$ (b) $d.A.V^2 Cos\theta$ (c) $d.A.V^2$ (d) None of these
5. Maximum efficiency of a series of vertical plates is
   (a) 66.67%   (b) 33.33%   (c) 50%   (d) 80%

6. The relation between hydraulic efficiency (η_h),mechanical efficiency (η_m) and overall efficiency (η_o)"
   (a) $\eta_h = \eta_o . \eta_m$ (b) $\eta_o = \eta_h . \eta_m$ (c) $\eta_o = \eta_m / \eta_h$ (d) None of these

7. A turbine is called impulse if the total energy at the inlet is
   (a) Kinetic energy   (b) only pressure energy   (c) sum of KE and PE   (d) None of these.
   1
8. Pelton Wheel is a

(a) reaction turbine   (b) radial flow turbine (c) impulse turbine   (d) None of these. 1

9. Francis turbine is a

(a) impulse turbine   (b) radial flow impulse turbine (c)axial flow turbine (d)reaction mixed flow turbine.

10. Kaplan Turbine is a

(a) impulse turbine   (b) radial flow impulse turbine (c) axial flow reaction turbine   (d)reaction mixed flow turbine.   1

11. Jet ratio (m) is defined as

(a) Diameter of jet of water to dia of Pelton Wheel   (b) Velocity of vane to that of jet   (c) Velocity of flow to that of jet   (d) Dia of Pelton wheel to that of jet of water.   1

12. Draft tube is used for discharging water from the exit of

(a) an impulse turbine (b) Pelton wheel   (c) reaction turbine   (d)None of these.   1

13.Main characteristic curves of a turbine means

(a) Curves at constant speed   (b) Curves at constant efficiency   (c) Curves at constant head   (d)None of these.

14. Operating characteristic curves means

(a)Curves at constant speed   (b) Curves at constant efficiency   (c) Curves at constant head   (d) None of these.   1

15.Muschel curves means

(a)Curves at constant speed   (b) Curves at constant efficiency   (c)curves at constant head   (d) None of these.   I   1

16. Governing of a turbine means under all working conditions

(a) head is kept constant   (b) speed is kept constant   (c) discharge is kept constant   (d)None of these.

17."To discharge large of liquid by multi-stage centrifugal pump, impellers are connected

(a) in parallel   (b)in series   (c)in parallel and in series(d)   None of these.

18. Cavitation will take place if the pressure of flowing fluid at any point is
(a) more than vapour pressure of the fluid (b)equal to vapour pressure of the fluid   (c)is less than the vapour pressure of the fluid (d) None of these.

19. Hydraulic coupling is a device used to
(a) transmit same torque   (b) transmit increased torque   (c)transmit decreased torque   (d)None of these.

20. Torque converter is a device to
(a) transmit same torque   (b)transmit increased torque   transmit
(c) decreased torque   (d)transmit increased or decreased torque.

21.Hydraulic ram is a device used for storing water in the form of
(a)pressure energy   (b)increase pressure intensity   (c)lifting small quantity of water (d)   None of these.

22. The flow of water, leaving the impeller in a centrifugal pump casing is
(a)Forced vortex flow (b)Free vortex flow   (c)Centrifugal flow (d)
        None of these.

Answers:
1(c), 2 (c), 3 (a), 4 (c), 5 (c) 6 (b), 7 (a), 8 (c), 9 (d), 10 (c), 11(d), 12 (c), 13 (c), 14 (a), 15(b),16 (b), 17 (a), 18 (c), 20 (a), 21(c), 22(b).

Bibliography

1. J. Twidell and T. Weir, Renewable Energy Resources, 2nd Edition,

2. O. Erdinc, M. Uzunoglu, Optimum design of hybrid renewable energy systems: Overview of different approaches, Renewable and Sustainable Energy Reviews 16 (2012) 1412– 1425.

3. Edgardo D. Castronuovoa, Joa~o A. Pec͵asLopesa,b, Optimal operation and hydro storage sizing of a wind–hydro power plant, Electrical Power and Energy Systems 26 (2004) 771–778.

4. Filipe Vieira, Helena M. Ramos, Optimization of operational planning for wind/hydro hybrid water supply systems, Elsevier: Renewable Energy 34 (2009) 928–936.

5. Alvin Toffler, 'Third wave'.

6. Renewable Energy Technologies: Cost Analysis Series - Hydropower, IRENA working paper, Issue 3/5, Jun 2012.

7. O. Paish, Micro-hydropower: status and prospects, pp 31-40, Proc. Instn. Of Mech. Engrs. Vol 216 Part A: J Power and Energy.

8. Hydropower in India: Key enablers of a better tomorrow, FICCI pwc publication available at https://www.pwc.in/en_IN/in/assets/pdfs/publications/2014/hydropower-in-india-key-enablers-for-better-tomorrow.pdf.

142

9. R.K. Rajput, 'A Text book of Hydraulic Machines', S. Chand & Company LTD Ram Nagar, New Delhi-110055, India.

10. Oliver Paish, Small hydro power: technology and current Status, Pergamon: Renewable and Sustainable Energy Reviews 6 (2002) 537–556.

11. Amity University Website, www.amity.edu/lucknow

12. Dr. R.K. Bansal, 'A Textbook of Fluid Mechanics and Hydraulic Machines', Laxmi Publications(P) Ltd, New Delhi, Boston USA.

13. NandKishor, R.P. Saini, S.P. Singh, A review on hydropower plant models and control, Elsevier:Renewable and Sustainable Energy Reviews, 11 (2007) 776–796.

14. V. Gupta, Dr. R. Khare and Dr. V. Prasad, Performance evaluation of Pelton turbine: A review, Hydro Nepal, Issue No.13, July 2013, pp 28-35.

15. A. Ruprecht, Unsteady flow simulation in hydraulic machinery, Task Quarterly, 6 No. 1(2002), pp 187-208.

16. MA Murtaza, Railway air brake simulation: an empirical approach, Proceedings of the Institution of Mechanical Engineers,
    Part F: Journal of Rail and Rapid Services.

17. MA Murtaza, SBL Garg, Transients during a railway air brake release demand, Proceedings of the Institution of Mechanical Engineers, Part F: Journal of Rail and Rapid Services.

18. R. K. Tyagi, Hydraulic Turbines and Effect of Different Parameters on output Power, European Journal of Applied Engineering and Scientific Research, 2012, 1 (4):179-184.

19. D. Agar, M. Rasi, On the use of a laboratory-scale Pelton wheel water turbine in renewable energy education, Renewable Energy 33 (2008) 1517–1522, Science Direct-Elsevier.

20. Loice Gudukeya,Ignatio Madanhire, Efficiency Improvement of Pelton Wheel and Cross flow turbines in Micro-Hydro Power Plants: Case study, International Journal Of Engineering And Computer Science ISSN:2319-7242Volume 2 Issue 2 Feb 2013 Page No. 416-432.

21. MA Murtaza & MI Murtaza, Hydro Power and Comparative Turbine Performance Evaluation',IJMPERD, Vol 5, Issue 6, Dec 2015,Page No. 43-52.

Other Books by the Author

1.Fundamental Concepts of Finite Element Method
(ISBN9798378375233)
   https://a.co/d/0gkHecK

[The book 'Fundamental Concepts of Finite Element Method' covers key topics like Engineering Design Process, Numerical Methods, and Applications of Finite Element Method. Tailored for undergraduates and postgraduates in Mechanical, Aerospace, and Civil Engineering, it addresses essential elements of FEM and is suitable for a broader audience interested in understanding the method.]

   2. Finite Element Method: Introduction, Concepts & Fundamentals:(ISBN9798363884245)
   https://a.co/d/bRR1MOa

['Finite Element Method: Introduction Concepts & Fundamentals' is a comprehensive guide covering CAD-CAE-CAM, Numerical Methods, and Applications of Finite Element Method. It is tailored to meet the requirements of undergraduates and postgraduates in Mechanical, Aerospace, and Civil Engineering, it's a valuable resource for those interested in FEM.]

3.Finite Element Method: Hands-on Introduction
(ISBN:9798854567770)

https://a.co/d/0xMzkUM

[This book, titled "Finite Element Method: Hands-on Introduction", covers essential topics for undergraduate (BE/ BS) and postgraduate (ME/ MS) students in Mechanical Engineering, Aerospace Engineering, and Civil Engineering. It guides readers through modelling techniques, programming languages, and the use of commercial software. The practical approach includes simple simulations using NISA and Ansys, aligning with the lab syllabus for Finite Element Method courses. The content is designed to help students gain proficiency in applying FE techniques to analysis and research, making it a valuable resource for anyone pursuing these engineering disciplines].

4. Finite Element Method: Theory & Practice (ISBN: 9798339838890)

https://a.co/d/hWsSYxs

[The book, "Finite Element Method: Theory & Practice", delves into Fundamentals, Formulation procedures and applications of the Finite Element Method. It caters to the syllabus of UG (BE/ BS) students in Mechanical/ Aerospace and Civil engineering, as well as PG (ME/ MS) students in Mechanical/ Automobile and Structural engineering. The comprehensive coverage ensures that those enthusiastic about studying FEM gain immense benefits from this book.

The worked examples and descriptions, enhancing the overall learning experience. The simplified language and clear explanations make this book an invariable resource for understanding Finite Element Method principles, applications, research and future trends].

About the Author

M. A. Murtaza, Professor Emeritus, Amity University (Lucknow Campus) India, received the B.S. degree from Harcourt Butler Technological Institute affiliated to Kanpur University. He received an M.S. degree in Design of Process Machines and Ph.D. from the Motilal Nehru Regional Engineering College Allahabad affiliated to Allahabad University. He started his career as a teaching faculty in the Department of Mechanical Engineering there. He served in Indian Railways as Director Research/ Research Designs & Standards Organisation Lucknow, Senior Professor in National Academy of Railways Vadodara and Chief Mechanical Engineer of Wheel and Axle Plant Bangalore. He re-entered in academics as Professor and taught M.S.(Design) and BS (Mechanical Engineering) at Oxford College of Engineering Bangalore and B.S. (Mechanical), B.S. (Aerospace Engineering) & M.S.(Automobile) at Amity School of Engineering & Technology Lucknow and was responsible for setting up of Centre of Excellence for FEM there. He has contributed three best papers and one of them was awarded 'TA Stewart Dyer Fredrik Harvey - Trevithick prizes' by I. Mech. E. (London).

https://in.linkedin.com/in/dr-m-a-murtaza-0239033

https://scholar.google.co.in/citations?hl=en&user=mGaaVIA AAAAJ

http://orcid.org/0000-0002-3958-8142)

Praise of Author

Dr. Tudor Rickards former Head/ HR of Manchester Business School, Manchester UK, has endorsed following recommendations on Linkedin.

"In his time at MBS, Dr Murtaza was an able and diligent researcher with a deep interest in management processes and creativity. He was also an excellent colleague to work with". (https://in.linkedin.com/in/dr-m-a-murtaza-0239033.

List of Symbols

N – Speed

P- Power

Q- Discharge

η- Efficiency

u – subscript for unit quantities,

s – subscript for specific speed.

o,m,h subscript for efficiencies.

V_r, V_f, V_w- Velocities, relative, flow & swirl.

Index

www.ingramcontent.com/pod-product-compliance
Lightning Source LLC
Chambersburg PA
CBHW071504220526
45472CB00003B/908